PARACELSUS'S THEORY OF EMBODIMENT: CONCEPTION AND GESTATION IN EARLY MODERN EUROPE

The Body, Gender and Culture

Series Editors: *Lynn Botelho*
Elizabeth Hurren

Titles in this Series

1 Courtly Indian Women in Late Imperial India
Angma Dey Jhala

Forthcoming Titles

Old Age and Disease in Early Modern Medicine
Daniel Schäfer

The Life of Madame Necker: Sin, Redemption and the Parisian Salon
Sonja Boon

The Prostitute's Body: Rewriting Prostitution in Victorian Britain
Nina Attwood

PARACELSUS'S THEORY OF EMBODIMENT: CONCEPTION AND GESTATION IN EARLY MODERN EUROPE

BY

Amy Eisen Cislo

Routledge
Taylor & Francis Group
LONDON AND NEW YORK

First published 2010 by Pickering & Chatto (Publishers) Limited

Published 2016 by Routledge
2 Park Square, Milton Park, Abingdon, Oxfordshire OX14 4RN
711 Third Avenue, New York, NY 10017, USA

First issued in paperback 2015

Routledge is an imprint of the Taylor & Francis Group, an informa business

BRITISH LIBRARY CATALOGUING IN PUBLICATION DATA

Cislo, Amy Eisen.
Paracelsus's theory of embodiment: conception and gestation in early modern Europe. – (The body, gender and culture)
1. Paracelsus, 1493–1541 – Influence. 2. Human anatomy – History – 16th century. 3. Medicine, Magic, mystic and spagiric.
I. Title II. Series
610.9'2-dc22

ISBN-13: 978-1-138-66382-4 (pbk)
ISBN-13: 978-1-85196-995-1 (hbk)

Typeset by Pickering & Chatto (Publishers) Limited

CONTENTS

ACKNOWLEDGEMENTS

I never thought that a three-page class paper would expand into a book, but that is exactly how this project began. In order to complete the requirements of the coursework towards my doctoral degree I reluctantly signed up for a class on witchcraft in Early Modern Germany. The instructor for that class, Gerhild Scholz Williams, asked us to consider the works of writers of the period whose texts were not included in the syllabus. I chose Paracelsus. Within a year I had expanded the paper and presented it at the Sixteenth Century Society Conference. Paracelsus scholar Allen Debus was in the audience and introduced himself after my talk. He encouraged me to stick with the project and that brief exchange led me to keep reading. I returned to the Sixteenth Century Society Conference almost every year thereafter and the feedback, encouragement and support I received from colleagues at those meetings was indispensable towards the final project. I owe my greatest thanks to Gerhild Scholz Williams for introducing me to the early modern period, introducing me to the Sixteenth Century Society and for advising the project when it was my dissertation

In addition to the wonderful feedback I received at conferences, I am grateful to colleagues who read and criticized sections of the manuscript in various forms or who otherwise gave me advice on particular points. The feedback and encouragement from the following colleagues made the project possible: Barbara Baumgartner, Lynn Botelho, Kathleen Crowther, Mary Ann Dzuback, Mary Fissell, Bruce Janacek, Christine Johnson, Bruce Moran, Michael J. Murphy, Linda Nicholson, Walton O. Schalick III, Jole Shakelford, Heinz Schott, Lynne Tatlock, and Andrew Weeks. I repeatedly returned to the works of two scholars I never met, but whose work helped me immensely in trying to figure out Paracelsus: Karl Sudhoff and Walter Pagel. My work would have been impossible were it not for the path that they, and other Paracelsus scholars, forged for me. I am also indebted to the staff of Washington University Libraries, especially the rare book area of Becker Medical Library. In particular, I wish to thank Lilla Vekerdy, who is now at the Smithsonian, for helping me locate every tract I wanted to examine. The many students who took my history of the body class at Washington University between 2007 and 2009 also deserve my thanks for asking good questions and for allowing me the opportunity to share with them the research that I did for Chapter 5. A great deal of thanks goes to Donna Kepley, the administrative assistant for the Women, Gender and Sexuality Studies programme, who helped me stay on top of all my teaching responsibilities during the process of completing this book. My friend Keri Cascio has put up with me for the entire duration of this project and I owe her many thanks. Finally, I wish to thank my family: the Eisens, the Cislos and host family, the Kopshoffs. I especially want to thank my husband, Geoffrey Cislo, and the people who have awakened me at all hours of the night, coloured on the drafts of these chapters, sighed heavily when they found me in my office writing early in the morning, laughed at my distracted attention and repeatedly told me, 'Just finish it already, Mom!'. It took me their entire lives to research and to write it, so it is to them that I dedicate this book: my daughters, Emily Nicole and Caroline Renée Cislo.

INTRODUCTION

> He was certainly a loud-mouthed and often drunk and boastful mystic; but he probably did much to deflect alchemy towards improving medical chemistry. His theoretical ideas were too clouded in mysticism to be useful, but in practical medicine he was more effective. He was one of the first to study occupational diseases and he recognized silicosis as a hazard for miners. He realized that goitre and creationism are related; and he used morphine, sulphur and lead in medicine, and mercury, with which he treated the then new disease of syphilis. He gave good descriptions of several types of mental disease, which he saw as an illness and not as due simply to demons. However, he firmly claimed that it is possible to create human life in the laboratory and gave full experimental detail on how to achieve this, starting with the fermentation of a sample of semen.
>
> 'Paracelsus (Lat.), Theophrastus Bombastus von Hohenheim (Ger.) (1493–1541)' *The Cambridge Dictionary of Scientists* (2002).[1]

The above excerpt gives a glimpse of the way Paracelsus's contribution to the field of medicine is usually assessed. Other reference sources call him 'the father of modern chemistry' or the 'father of modern pharmacology'. His character flaws: drunkenness and abrasiveness, usually serve as qualifiers to his absurd contention that he could create human life in a laboratory. There are two major problems with this assessment. On the one hand, his drunkenness was documented by a student of his named Johannes Oporinus (1507–68) which raises the question of how accurately the student described his mentor.[2] On the other hand, the authenticity of the tract on creating human life (*Liber de humonculus*) has been questioned by compliers of Paracelsus's works.[3] Contrary to popular belief, Paracelsus firmly believed in the sanctity of human life and spent many years contemplating the degree to which God played a role in forming human bodies.

The bizarre and inaccurate descriptions of Paracelsus did not begin in the modern era, but rather are part of a legacy tied to his quest to define himself as a new and different kind of doctor. He vehemently opposed scholastic medicine that relied on the teachings of Aristotle and instead focused on observing nature and reflecting on Biblical teachings to understand the human body. His detractors, like Oporinus, T. Erastus and others, never cared for his approach to the body and, as the above passage makes clear, they seem to have succeeded in marring his name and his profound influence on modern medicine.[4]

Because I began reading Paracelsus's writing long before I ever consulted a reference work that provided a condemnation of his character, I encountered a Paracelsus who was profoundly interested in understanding the meaning of human bodies and how (or if) the physical body related to the divine. Paracelsus proclaimed his place as a revolutionary in the field of medicine, and in particular in the field of reproductive medicine, even though he is most often associated with chemistry. While his influence in chemical-based medicine has received a great deal of attention, I wondered why Paracelsus's interest in understanding conception and gestation was limited to the myth of creating life in a laboratory. He writes extensively about human birth in his vast oeuvre, but this aspect of his work is scarcely mentioned. To me his theories of human birth seemed intricately tied to his chemical approach to medicine, his understanding of the patient, his quest to define himself as a new and radically different kind of doctor, and his philosophical quest to understand the meaning of the human body. Why would histories of medicine and science not investigate his theory of understanding the human body if he was advocating a new approach to medicine? To answer this question, I began combing the scholarship on Paracelsus.

I learned a great deal about how Paracelsus understood the chemical balances of the body based on mercury, sulphur and salt and that he supposedly made connections between the body and astrological powers. However, the studies I initially consulted tended to gloss over how he understood the human body and reduced his understanding to a contemplation of the parallels between the human body as a microcosm of the larger world (the macrocosm). I suspected that Paracelsus's chemical philosophy altered the way physicians thought about the physical body. Yet what I learned from Allen Debus's research was that Paracelsus's chemical philosophy changed the way scientists performed laboratory work and that Paracelsus used medieval alchemical philosophy as the basis of his highly influential chemical philosophy.[5] The work of Walter Pagel and Allen Debus solidified the idea that Paracelsus revolutionized science by changing laboratory methods and introducing chemically-based pharmaceuticals. Earlier research (that is before 1980) on Paracelsus articulated his place in the field of medicine and specifically chemical medicine, but it did not tie in his theological interests. I realized that the scholarship was highly dependent on the way Paracelsus's writing had been collected. Until recently, only the medical tracts were collated and most of the theological writing remained difficult to find. It is no wonder then that early research focused only on Paracelsus's medical and chemical philosophy.[6]

When more of Paracelsus's oeuvre became available to us and people realized that his theological interests were vast, the research topics likewise expanded. During this time period feminist scholars investigated Paracelsus's theological writing and pointed out some of the more complimentary ways that Paracelsus wrote about women in comparison with many of his sixteenth-century contemporaries. Generally, these studies focused on the social status of women as revealed in his theological writing and therefore were not necessarily concerned with the female body.[7] However, in the context of witchcraft and understanding the biblical figure Mary, Paracelsus's writing on the composition of the body gained some attention.[8] Paracelsus opposed the idea that women became witches because of a contract with the devil (a widespread assumption at the time) and instead argued that evil entered the body at the moment of birth, similar to the way illness or disease entered the body.[9] Feminist scholars concluded that Paracelsus was an early advocate for women. The question of how Paracelsus interpreted women's social roles seems to have been answered in these studies, but I got the sense that there was more to how Paracelsus understood the female body.[10] However, research turned away from the feminist inquiry I just described and demonstrated the need to understand Paracelsus's writing in the context of his overlapping interest in medicine and theology. In these works I found recognition of Paracelsus's theory of embodiment: the idea that humans consist of two bodies, one physical and one spiritual, but no discussion of the practical ramifications of understanding the physical body as sexed.[11]

My assessment of secondary literature on Paracelsus made clear that research has not been specific about how this theory of embodiment emerged. In addition, scholarship on Paracelsus's understanding of the body overlooked an aspect that I found central to Paracelsus's writing, namely how it grew out of his desire to explain conception. Furthermore, it seemed obvious to me that Paracelsus believed that recognition of sex difference was necessary for proper medical treatment of the patient. I could not figure out why this important aspect of Paracelsus's medical theory lacked recognition. Only after years of research did I stumbled upon an article written in the 1960s by one of the foremost Paracelsus scholars, Walter Pagel. He wrote: 'A principle fundamental to Paracelsus' view of the world and man is that of *sexual duality*. It is a principle that he finds operative even in disease. Epilepsy in the female is a different disease from epilepsy in the male. A drug effective in men, may fail in women'.[12] Much to my dismay, Pagel's recognition of sex-differentiation was left out of future histories of Paracelsus.[13] This was a topic that I needed to research further. I wondered how Paracelsus came to the conclusion that humans consisted of two bodies and how he decided that men and women should receive different sex-specific medical treatment. Since I did not find detailed discussion of this topic in research devoted to Paracelsus, I turned my attention to scholarship on the history of the body and history of women's health and found that even there Paracelsus was barely noticed.

I knew that Paracelsus invited his readers to consider that some aspects of the body were physical and natural, while other aspects of the body were invisible and divine. He also insisted that physicians recognize the core difference of male and female bodies and treat patients accordingly. He focused on the significance of the female body, not as a lesser version of a male body; but rather as deeply connected to the divine realm. Paracelsus's profound analysis of the female body and the idea of a natural body separate from the spiritual self did not seem to resonate with those who aimed to understand the history of the body. For example, Thomas Laqueur declared that, 'Sometime in the eighteenth century, sex as we know it was invented'.[14] He explained that prior to the eighteenth century, people thought of male and female bodies as versions of each other: the sex organs of women were inverted versions of the sex organs of man; and the way sex was represented and understood was primarily influenced by social expectations. Before the mid-eighteenth century, he argued, there was but one sex and that was male, the female was simply a lesser version. What changed in the eighteenth century, Laqueur argued, was that sex organs themselves began to signify social status, the idea of natural law emerged from the notion that one's sexed-body determined ability. Unlike earlier periods the evidence in nature was perceived to dictate an idea that male and female were not versions of each other, but rather different sexes entirely. Sex as we know it emerged in the eighteenth century? But, Paracelsus did evaluate 'natural' difference between men and women. Furthermore, I was under the impression that modern understanding of sex difference rested, in part, on understanding hormonal differences. Paracelsus insisted on chemically-based difference of male and female bodies, a point that did not enter into Laqueur's assessment of the sixteenth century. How could the 'father of modern pharmacology' not matter? This is one question my book seeks to answer, but I soon realized that I was not alone in doubting the time frame for the emergence of a two-sex body.

Historians of medieval and early modern Europe were sceptical of Laqueur's claim. Because he focused heavily on anatomy, they focused on visual representations of the body and the act of dissection to disprove the idea that the two-sex body was an invention of the eighteenth century.[15] At first, most of the research on the history of the body in medieval and early modern Europe focused on the writings of university-educated physicians. The source material was therefore written in Latin for the educated class.

Because Paracelsus wrote in German, rather than Latin, I was drawn to research on vernacular medical texts and particularly those focusing on human birth. The work of Mary Fissell caught my attention. She explored English vernacular texts and articulated the importance of studying how people wrote about female bodies as 'maps of gender relations'. Fissell argued that men's writing about women's bodies in the latter half of seventeenth-century England used the female body to articulate gendered hierarchical ideals. She studied low-cost vernacular (English) prints concerning human reproduction and explained how writing about reproduction was widespread in England in the second half of the sixteenth century. In arguing that perceptions of the body were local, she distinguished the viewpoint of the English from Continental gender expectations based on the writing of Thomas Raynalde. She explained how Raynalde adapted the German midwifery text by Eucharius Rösslin to make it more sympathetic to female practitioners of medicine and to honour the role of women in childbirth. Fissell also drew on the sermons of exiled Anabaptists, like Martin Bucer, to explain the Continental influence on thinking about Mary's role in reproduction and reproduction in general in England. Fissell invited us to consider how the Reformation in England instigated a change in the way people thought about reproduction. [16] Her research draws on the transformation of German texts, which leaves open the question of how reproduction was understood in Germany and how the body's sex was articulated in German vernacular texts. Fissell's astute observation about the local nature of how bodies were understood begs the reconsideration of Paracelsus's writings on the topic. Paracelsus influenced future writing on the workings of the female body that resonated in the works consulted by writers in Reformation England.

Paracelsus's writing on embodiment that emerges from his theories of conception and gestation broadens our perspective on how the body was interpreted on the European continent from a different perspective. Paracelsus played a significant role in advocating a German approach to medicine.[17] Even though research has shown that the people of early periods represented the physical body of women as anatomically different to men's bodies and that women's bodies were understood in different physical ways, we still lack information on how people theorized the body in the early modern Germany (1500–1700). What I mean by this is that the scholarship available interprets the way people read the physical body in medieval and early modern Europe, but we lack a first person account that reveals an intellectual engagement with understanding the body.

The writings of Paracelsus offer a glimpse into this important aspect of understanding human bodies in early modern Europe and helps explain the transition in thinking that Laqueur assumed began in the eighteenth century and that Fissell recognizes in British vernacular texts on human birth. The influence of Paracelsus in this field has been undervalued. I argue that the idea that male and female bodies were quantifiably different emerged in the writings of Paracelsus and his followers helped usher in the idea of a chemically-based difference in the sexes. Paracelsus's writing demonstrates that it was not empirically-based science that solidified this idea, but rather his contemplation of the meaning of the physical body through comparison to his understanding of the bodies of holy Christian biblical figures: Christ, Mary, Adam and Eve. Furthermore, Paracelsus's quest to understand the physical body grew out of his desire to carve out a niche for himself in the ever-expanding field of medicine. Current scholarship demonstrates that discourses about human birth (conception and gestation) often anchored discussions about the meaning of the body and of the soul in Reformation Germany, although most references to this phenomenon cite works composed after Paracelsus's writing.[18] Paracelsus may have contributed to the widespread discussion about human birth and the origins of the soul. He develops a theory of embodiment, the idea of the physical body and soul as separate, in the course of thinking through birth.

This book explains Paracelsus's philosophy of the sexed-body and shows how his ideas were disseminated by his followers in the latter half of the sixteenth century to the end of the seventeenth century. Through close readings of several of his tracts on conception and gestation, I show how his theory of the natural body as separate from the spiritual self emerged. Paracelsus defined the human body as partially a product of the natural world, with nature understood as an entity separate from the divine realm. He came to this conclusion after many years of contemplating divine involvement in human conception.

Current writing on the topic of how we understand sex difference still assumes a parallel development between the emergence of empirical science, the foundations of natural law and the idea that there are only two sexes. For example, Anne Fausto-Sterling implores us to recognize the way we have socially constructed a sex-binary based on the false belief that human biology is limited to two sexes. She argues that the contemporary scientific approach to the body has led to widespread cultural assumptions that sex, gender and sexuality can be understood on a microscopic or chromosomal level. Throughout her book she articulates that this understanding of the body is a thoroughly modern one that is based on quantifiable evidence-based science and/or medicine.[19] Fausto-Sterling's research does not in any way contradict Laqueur's statement because she focuses entirely on twentieth-century science. However, her book underscores the idea that the modern concept of the two-sex body, also called the sex binary, is based on the notion that science provides us with indisputable facts or quantifiable evidence that we use to reinforce the social idea of man and woman. In the twentieth century, Fausto-Sterling argues, hormones (i.e. chemical difference) took centre stage in research aimed at understanding sex-difference and homosexuality. This study makes clear that such thinking existed as early as the sixteenth century, even if the terminology differed.

The writings of Paracelsus reveal how this social construction of scientific facts about sex is rooted in Christian theology and how it emerged not from the highest medical authority of the day, but rather from medical writers seeking the approval of general (possibly uneducated) readers. It also demonstrates that Paracelsus considered that the body's sex could be understood on a microcosmic scale.[20] Although Paracelsus did not have the technology or the vocabulary to explain hormonal or chromosomal differences, he did, in fact, think about the articulation of sex as a natural physiological difference. His writing reveals that it was not laboratory experiments or hands-on dissection that led to his conclusions, but rather thinking through the meanings of the human body in comparison and in contrast to biblical figures and thinking through the limits of divine acts on the human body.

We know that Paracelsus played an important role in the development of pharmaceutical medicine, but there is much more to the story of how Paracelsus influenced the future of medical and popular discourse concerning the body and the body's sex. His writing reveals the emergence of theorizing the body as a scientific object, his personal struggles to come to terms with his religious ideas about the physical body and his suspicion that some aspects of the human body may not be divinely willed. Understanding Paracelsus's theory of embodiment, or the way in which he concluded that humans have both divine and natural qualities within them, clarifies a period of medical intellectual thought that has heretofore not gained enough recognition. It is time to consider the way that Paracelsus engaged with the topic of human birth, rather than to perpetuate the myth of Paracelsus as the drunken, loud-mouthed, mad scientist who concocted human life in a laboratory.

1 RETHINKING SEX IN EARLY MODERN GERMANY: NEGOTIATING MEDICAL AUTHORITY

Paracelsus's earliest writings reflect his desire to situate himself in the medical profession. Within these texts he repeatedly turns to the topic of conception and the workings of the female body. He was not the first physician to attempt to carve out his niche in the profession by asserting his expertise on the topic of human birth. Monica Green explained this phenomenon as beginning in the medieval period, in other words more than 100 years before Paracelsus began writing.[1] But Paracelsus's writing takes on nuances of the sixteenth century as he incorporates concerns of his day into his assertion that he understands the process of human birth better than his competitors. He does not, however, abandon approaches to the understanding the body that were prevalent in medieval medical and theological exposition. He asks himself how the womb works and how the material necessary for the development of new life converges. He ponders the moment when a physical body is formed and wonders if it occurs at the same moment one gains a soul. In this way, he still grapples with questions posed by medieval intellectuals.[2] However, unlike his medieval predecessors in the German-speaking regions, Paracelsus wrote during a period of ever-expanding authority in the field of medicine. He found himself competing with university-educated physicians, midwives and the vast array of practitioners who also sought to assert their authority in the field of gynaecology.

It was a time in which concern for the unborn and the fear of witches and monsters prevailed. People sought answers to what caused unusual or monstrous births and they became suspicious of women caring for women in the birthing room for fear of satanic intervention at such a precarious time. Paracelsus was aware of these concerns when he began writing and he sought an alternative explanation for his own authority and for explaining the process of conception.

His argumentative style aims to persuade readers to give up the way classical authorities wished to explain the process of conception and instead follow his model of conception based on scripture. In the process of articulating his theory of conception, Paracelsus rethinks the term sex in its two modern usages. On the one hand, he reflects on the process of conception, all the while avoiding writing about the act of coitus and instead focusing on the reproductive material. On the other hand, he ponders the meaning of human physicality and how it relates to Christ as he begins to articulate an idea of the sexed body, i.e. male and female. Paracelsus's desire to carve out a niche for himself within the ever-expanding field of medicine in sixteenth-century Germany influences his philosophical quest to understand the process of conception and the meaning of human physicality. He begins to participate in a discourse of using an understanding of the workings of the body, and particularly the female body, as a way to captivate an audience and strengthen his authority.[3] While Paracelsus ultimately arrives at a theory of human physicality that takes into account a two-sex body (distinct male or female qualities), he begins his query by attempting to understand the process of conception and gestation. This chapter serves to orient readers to the climate in which Paracelsus wrote and should provide context for the close examination of Paracelsus's writing that shall be addressed in Chapters 2 through 4.

The Medieval Tradition

Paracelsus's contemplation of human birth builds upon medieval alchemical, theological and medical reasoning. Paracelsus is well-known as an alchemist, a field that gained momentum in the medieval period. A core principle of medieval alchemy was the philosopher's stone and it was widely believed this stone had the ability to transform common metals into gold. Indeed, alchemists wrote about this transformation as a rebirth, and the philosopher's stone was described as a union of male and female chemical bodies that gave birth to a hermaphrodite.[4] Paracelsus approaches human bodies in much the same way as medieval alchemists explained the transformation of metals through the philosopher's stone. He also employs a kind of alchemy to understanding the process of conception when he thinks about the union of male and female seeds. Like alchemical writers before him, he declares his opposition to an Aristotelian theory of generation in which all parts of the existing mixture are the same as the whole.[5] While he agrees with the chemical principles of medieval alchemy and the importance of mercury and sulphur, he extends such thinking to consider the formation of human bodies. He insists that human bodies are formed through a chemical process that unites male and female seed; but he ponders the significance of Christ's birth in trying to understand human physicality. In this way, Paracelsus was participating in what scholars tend to demarcate as a medieval tradition, or discourses about the body before 1500. We think that he began writing in the 1520s, and it would be all too simplistic to assume that the ideas of those writing in the Reformation would abruptly deviate from medieval alchemical, theological or medical theories. Paracelsus incorporates, yet questions, medieval ideas about the body as he asserts his authority in the early modern world.

The way in which he approaches his inquiry by contemplating the bodies of biblical figures draws on medieval understandings of Christ's body. Medieval writers represented God's body with both feminine and masculine characteristics. For example, when people wrote of Christ's blood in the medieval period they contemplated the workings of his body much like the workings of a female body and understood his bleeding as a menstrual process, as Caroline Bynum has shown.[6] Bynum further demonstrated that Christ's body was interpreted as neither male nor female, but rather as an androgynous being. Such a construct of the holy body of Christ was not altogether different from the way medical texts portrayed the human body in the medieval period. Medical writers assumed that the treatment of male and female patients was generally the same and anatomical illustrations of the time did not differentiate male and female skeletal structures. In Paracelsus's writing the connection between the medieval concept of Christ's body, human bodies and alchemical seed theory are all questioned.

Paracelsus's theories of human and holy bodies that he articulates in the process of explaining human conception demand our attention because he represents a period that has been misunderstood in accounts of history of the western philosophy of the body. In trying to understand how we have come to explain the body as consisting of two sexes in the modern day, scholars regard the modern perception of a sex-binary as a scientific convention and they seek evidence of this shift in anatomical models. Thomas Laqueur called the androgynous model of the body described above a 'one-sex body', although such thinking did not prevail until the eighteenth century as he claimed.[7] Is it important to recognize how thinking about the body as a 'two-sex' model, to use Laqueur's term, grew out of theological contemplation of the body, not solely from medical or scientific inquiry. Paracelsus attempts to understand the human body by comparing its formation to that of Christ, Mary, Adam and Eve. He moves away from the medieval tradition of defining Christ's body in terms of a combination of male and female, but still builds on a medieval alchemical model of uniting chemical properties when he develops his understanding of human conception. Paracelsus incorporates concerns about the body that were specific to his time and place, but also grapples with medieval constructs of understanding the body. In this way, Paracelsus's work represents a crossroads between medieval and early modern thinking about the meaning of the physical body. His response to contemporary concerns about the body and about human birth catalyse his alternative explanation for the process of human conception and his rereading of the meaning of the physical body.

Scrutiny of the Body in Early Modern Germany[8]

Paracelsus wrote in German and for the most part lived in German-speaking regions. He was born in Einsiedlen (currently located in Switzerland)[9] and spent time in Basel, where he became city physician in 1527 after he had treated the city's most famous publisher, Johannes Froben. Froben was so pleased with Paracelsus's skill that he managed to have him appointed to the university in Basel.[10] His appointment in Basel ended abruptly when he was forced to leave because of his unconventional teaching (he denigrated the classical medical texts) and he then travelled throughout German-speaking regions.[11]

Paracelsus was undoubtedly familiar, though not necessarily conscious of, the multiple discourses about the female body in his day. Representations of the New World in visual media used the naked woman as a symbol of the fertility of the newfound lands.[12] Yet, beyond the female body as emblem, as Katherine Park has called it, and aside from using the female body as a discourse, there was also widespread concern about the workings of the female body in German-speaking regions.[13] In Early Modern Germany (the period beginning around 1500) certain areas witnessed an intense scrutiny of possible cases of witchcraft. It is well known that during witch trials women were questioned about their fertility, the number of living children they had and their experiences during the lying-in period.[14] In this way, the topic of conception, gestation and birth occupied the thoughts of those worried about witches. Paracelsus did not share the opinion of those who believed that women were more easily swayed by the devil and denied the widespread belief that women entered a pact with the devil through coitus.[15] Although Paracelsus briefly enters into discussions about the root cause of a person becoming a witch (he denies the widespread belief that a pact is made by having sex with the devil); he focuses in on the process of conception and gestation, a topic of great concern to those worried about possible sorcery. Because the lying-in period was considered a moment when women were most susceptible to the devil, people attempted to protect women during this time. Moreover, in sixteenth-century Germany the pregnant body held special status. Legal documents show the ways in which pregnant women gained protection either for accommodations during the lying-in period, the right to beg for money, or the right to nourishment.[16] The protective measures and sanctity of the pregnant body arouse from Protestant teachings about the importance of child-bearing, but the concern for witchcraft also led people to want to protect pregnant women and to scrutinize the female body.[17] Concern for witches drew attention to pregnancy and understanding the workings of the female body. But there were other interests concerning birth in the German-speaking world of Paracelsus's day: people wondered about what we would today call birth defects, in those days people spoke of 'monsters'.

Growing interest in abnormal births was documented on broadsides in early sixteenth-century Germany. For example, at the end of the fifteenth century in Basel, Sebastian Brant published a number of pamphlets depicting unusual births. In these pamphlets he also wrote about the possible meanings of the birth, noting that it could be a natural change instigated by God as a warning to humans. [18] The birth of the famous 'monster of Ravena' in Italy in 1512 fuelled discussions about the meanings of the occurrence, as well as the causes.[19] From Protestant discussions about the importance of marriage, to the fear of witches and monstrous births, concerns about births were widespread in sixteenth-century Germany. The representation of unusual births elicited a range of responses from awe to fear, and from wonder to disgust. At the same time associations of the female body to the process of birth fostered questions about women's possible power in changing the body of the baby. Ultimately, questions about unusual births placed the workings of the womb into the forefront of discussions of power. Keeping in mind the concern for sorcery, there was a need to understand the process of conception to preclude the idea that a birth was satanic and, in this regard, Paracelsus sought to voice his opinion on the process of conception and the meaning of the sexed body. However, he may have had another agenda: he needed to secure his authority as a physician in order to continue to profit from his medical advice. There were competing theories of how conception occurred, and it is this aspect of understanding both male and female bodies that interested the young Paracelsus.

Theories of Conception

Paracelsus develops his theory of conception by strongly opposing the reliance on classical texts that many of his contemporaries used. There were, in sixteenth-century German print, a variety of theories of how conception occurred. Some adhered to an Aristotelian idea that only the male had seed and the female served as a receptacle for that seed. Others drew on the teachings of Hippocrates and Galen who posited that men and women each produce seed that forms the foetus. The Hippocratic model followed the idea that the strength and quantity of the seed contributed to the appearance of the offspring. In Hippocratic theory each part of the body of each parent contributes some aspect of itself to the seed. The Galenic model relied on the humours, in which females are considered cooler and for that reason their seed was less important than the male seed.[20] The teachings of each of these classical authors appeared in German-speaking print, while medical education of the time required familiarity with Aristotle's theory. [21] For university-educated theorists Aristotle's teachings found most favour because his works were required to fulfil the medical degree and while those with university degrees wrote in Latin, discussion about theories of conception also occurred in vernacular (German) texts.

There were multiple and competing expositions on the topic of conception and gestation circulating in both manuscript and printed form. Paracelsus aimed to participate in the vernacular discussion of the process of conception as a way to assert his medical authority, but he met resistance. A popular medical writer of the time, Walther Hermann Ryff (1500–48), castigated Paracelsus's expertise on the topic of conception by framing him as corrupt:

> Theophrastus made a special herb that has power to make a person infertile. Our pregnant virgins have used its special power so that they can carry on their secret whoring and not conceive.
>
> *Theophrasto wird ein sonderlichs Kraut gestezt/ welches die Kraft und Eigenschaft haben sol/ unfruchtbar zu machen. Unsere schwangere Jungfrauen haben auch ire besonderen kraften damit sie ihre hürerei heimlich treiben mögen/ und nicht entpfahnen.*[22]

While the passage could refer to the classical writer, Theophrastus, most likely it was an attempt to take a stab at Paracelsus, who at the time went by the name Theophrastus.[23] Either way, Paracelsus was leading an uphill battle in trying to define his authority as a physician. Ryff does his best to cast an ominous light on Paracelsus's approach, which as we shall see was gaining favour. Ryff and other writers sought to underscore their own medical authority by highlighting their expertise in the field of women's reproductive health. They opposed the authority of midwives, who they saw as their primary competition.

Books that denigrated the role of midwives and elevated the status of male physicians had become popular in sixteenth-century Germany and translations of these texts spread throughout Europe. Perhaps the most significant text on midwives and their practice, in the context of this study, is Eucharius Rösslin's *Der Schwangern Frauen und Hebammen Rosengarten*, published in the vernacular under permission granted by Maximilian in 1513. Rösslin's book provided information for pregnant women and cautioned against unskilled midwives. This enormously popular book described, in German, the process of birth and detailed ways to help a woman in labour. It was reprinted several times in German and then was translated into English twice, the second translation, by Thomas Raynold, was the most popular.[24] Unlike Paracelsus's highly philosophical writing on the topic, Rösslin's book was a sort of practical guidebook to women's obstetrical care and infant paediatric care. Rösslin's book does not include an explanation for the process of conception as does Paracelsus's; instead, the book focuses on how to understand the pregnant female body. Rösslin repeatedly cites the authority of classical authors whose work was acceptable to university-educated physicians. Rösslin favoured the authority of physicians educated in classical texts and emphasized such theories in his book, however the foremost purpose of his book was to caution against the poor skill of uneducated midwives.[25] Walter Herman Ryff's *Frauen Rosengarten* followed the publication of

Rösslin's. Ryff's work went through several editions in the early sixteenth century.[26] Although the dating of Paracelsus's early works is imprecise, it is likely that their texts on the topic of conception were composed around the same time. Ryff was a contemporary of Paracelsus, though he was a different kind of doctor as revealed among other things, in their varying approaches to the topic of conception.[27] Ryff considers how conception occurs and bases his understanding on classical texts. He also implores his readers to visualize the baby inside the mothers' body. Using line drawings alongside the narrative, he attempts to define the process of birth citing the authority of the famous medical philosophers Albertus Magnus and Avicenna. When referring to Aristotle, Ryff focuses on lack of heat and cautions men to avoid being in ice-cold water because this chills the seed so that it will not function.[28] Although Ryff does mention possible causes of infertility, his most emphatic warning emerges in the segment on the evil ways that some people murder a baby while still in the womb. Ryff warns against midwives and other unscrupulous physicians, namely Paracelsus. Paracelsus never targets midwives, but instead condemns people like Ryff. He warned against using classical authority to understand the process and, in contrast to Ryff, he never uses illustrations in his explanations.

In most regards, Paracelsus's approach to asserting his authority in the field of medicine, particularly gynaecology, sharply differed from the ranting diatribes against midwives favoured by Rösslin and Ryff. However, Paracelsus does ascribe to using metaphors of the Fall to explain human physicality and this can also be found in the work of Ryff and another popular writer, Horst. Research by Kathleen Crowther demonstrates the popularity of authors Walther Hermann Ryff (1500–48) and Jakob Horst (1537–1600). Crowther explains how vernacular authors frequently wrote of conception using metaphors of the fall in Genesis and how many envisioned God directly shaping the contents of the mother's womb. Ryff, in most instances, described human conception using a Galenic model, in which the seeds of man and woman combine to form offspring. The writings of Horst and Ryff make clear the interplay between devotional teachings, religious beliefs and theories of conception. Paracelsus, who wrote most of his tracts on human birth before Horst, contributed to this aspect of the widespread interests in conception by rejecting Aristotle and emphasizing biblical sources of understanding.[29]

Paracelsus's decision to focus on biblical authority to assert his own expertise may have stemmed from Ryff's slander, or he may have coopted topics of interest at the time. In the early sixteenth century, around the same time Paracelsus began writing, Cornelius Agrippa of Nettesheim wrote of the need to understand the equal nature of men and women based on the biblical model of Adam and Eve in his tract, *De nobilitate et praecellentia foeminei sexus* of 1529.[30] Among protestant writers of the time, the Genesis story served to underscore both women's equality with men and women's inferiority. Luther, for example, used the story of Adam and Eve in various ways; but in referring to the workings of the female body he tended to regard her reproductive role as the reason for her submission.[31] While on the one hand, discussions about women's nature entailed a comparison to the importance of women to Eve, people tended to read the nature of women as sinful, based on the idea that Eve's weaker mentality made her more prone to persuasion. While some Protestant writers wrote about the female body as a way to explain or reinforce the social status of women, Paracelsus took the discussion in an entirely different direction. He interweaved the story of Genesis and a theory of conception as he pondered the meaning of the physical body. Paracelsus was writing while these competing ideas about conception and the nature of women were hotly contested. He participated in theories of conception, but tended to refute the Reformer's claim that there was something inherently weaker about the female body. In contrast to Reformers, like Luther, who used the divinely-willed weakness of the female body to substantiate reasons for women's submissive status, Paracelsus uses the topic of conception and the workings of the female body to assert his own authority as a physician. While other writers of his day asserted their authority using metaphors of the Fall combined with classical theories of conception to assert their prominence and castigate midwives, Paracelsus draws his battle lines differently. At times he employs metaphors of the Fall, but in his early works he chooses to compare human birth to Christ and he does not address the issue of midwives, but instead focuses heavily on Jewish doctors.

Perhaps conscious of the similarities of his own approach to medicine with Jewish practitioners in German regions, Paracelsus devised a way to assert his authority that would resonate among his readers. He turned his animosity toward Jewish physicians who were finding it increasingly difficult to practice in the area.[32] In fact, they were banned from German-speaking regions as early as the late fifteenth century.[33] Paracelsus explicitly condemns non-Christian healers throughout his works and it is therefore important to consider his rhetoric within the context of challenges within the medical profession and within the context of Protestant writing against Jews.[34] Until recently, little information was available on Jewish physicians in Paracelsus's time, particularly on those in Germany. Generally, scholarship on the history of medicine focused on Jewish doctors in Spain and Southern France. In these areas, Jews were considered well-educated because they had access to Arabic texts.[35] Research on Anthonius Margaritha's texts represent the kind of writing that condemned Jewish physicians in German-speaking regions in Paraclesus's day. Margaritha, the son of a rabbi, converted to Christianity in 1522. His *Ganz Jüdish Glaub* was published in Augusburg in 1530. Michael Walton demonstrates how Margaritha condemned Jewish physicians because of their lack of education (Walton notes that they were excluded from German universities at this time), their inability to use Latin, and their remedies that included mercury and sulphur.[36] Walton concludes that Margaritha's observations about the poorer medical practice of Jews in German-speaking regions may not have been too farfetched. He explains that Margaritha spent time in Italy, where Jews were not excluded from medical education and that his experience allowed him to compare Jewish doctors in Germany who did not have access to formal education, to those in Italy who did have such access. Margaritha's comment that the Jewish physicians prepared pharmaceuticals using mercury and sulphur is important to this analysis because these are two chemicals that Paracelsus frequently discusses as cures for ailments. An interesting dynamic seems to be at play here. On the one hand, Paracelsus writes in the vernacular and it is possible that he, like the Jewish physicians Margaritha condemns, was unskilled in Latin. On the other, Paracelsus may have seen the need to speak out against those who were profiting using the same techniques he advocated. There is still another area to consider in understanding Paracelsus's stance against Jewish physicians. In the 1540s the well-known reformer Martin Luther warned against Jewish doctors.[37] It is possible, at least in Paracelsus's later writings, that he was using a rhetoric that was fashionable at the time in order to strengthen his own position as a reformer of medicine.[38] Considering the way Paracelsus invents his own definition of doctoring as an opposition to the existing practices opens a means for understanding how his theory of embodiment grew out of his stance as a new and different kind of doctor. Before turning to his early theory of conception, that includes his self-definition, it must be noted that Paracelsus's approach to medicine was severely threatened and this too significantly influenced why he wrote extensively about conception and gestation. On the one hand, he was engaging in a topic of widespread interest that drew on theories of witchcraft and discussions of deformed births. On the other hand, he was arguing for his authority as a physician.[39]

With the rise in university-educated doctors, the countermovement of medical practitioners consisted of many Jewish doctors. Paracelsus sought to distinguish and defend himself as a means to assert his authority as a physician. The advancement of German medical education necessitated his stance. The first university in a German-speaking region was founded in Prague in 1348, followed by Vienna in 1384. The universities in the German-speaking areas where Paracelsus is believed to have spent some of his early years were founded considerably later. For example, Basel was founded in 1460.[40] These universities offered a full range of academic subjects, including medicine, but the number of students who graduated with a medical degree from these universities in Paracelsus's lifetime was very small. Most universities lacked a specific area for medical education, and masters taught the small number of students interested in medicine in their homes.[41] The material taught in the German-speaking universities relied on translations of standard authors of the Greco-Arab canon, but there is evidence to suggest that medical students also read somewhat newer commentaries on medicine by French and Italian physicians.[42] In the German areas understanding the body was still very much a philosophical endeavour that involved little hands-on exploration. Whereas the medical programmes in France and Italy included dissection as part of the curriculum, this was not the case in Germany. Scholars have generally concluded that knowledge of anatomy was limited in the German regions.[43] Even though faculty and students petitioned to have human dissections taught in the medical school of Wittenberg university as early as 1502, dissections were first established as part of the curriculum in 1511, but even then they were scheduled to be offered only once every three years.[44] Generally speaking, the type of medical education that students received in German-speaking universities during Paracelsus's lifetime was primarily theoretical; practical training was extremely limited. Indeed, the primary focus of universities in German-speaking regions was Theology and Law; few universities had programmes in Medicine.[45] Using the example of Basel, where Paracelsus lectured and lived, it took years before they moved away from the classical medical standards reflected in Galenic teachings. It was not until late in Paracelsus's life that anatomy was officially included in German medical education. Sixteen years after Paracelsus lectured in Basel, the Dutch physician Andreas Vesalius carried on the tradition of questioning Galen and emphasized the importance of anatomy in medical education.[46] His *Fabrica* (1543), a 700-page tome illustrated with many anatomical images, altered the course of medical education in general and especially in the German-speaking world.[47] Since we have evidence that anatomy was introduced in the curriculum of Basel University after Paracelsus departed, we do not know if he ever witnessed a dissection. He may, however, have encountered anatomy discussions or even a dissection during his travels. His writing suggests that he remains very rooted in the kind of medical education available to those in German-speaking regions. In other words, he explores the workings of the body from a philosophical standpoint and does not try to reconcile his thinking with any kind of anatomical observation.

Paracelsus's medical interests were deeply intertwined with his desire to maintain his own authority. Even though there is some chance that he may have benefited from studying anatomy in Italy, he remained deeply rooted in the sentiments of embodiment expressed in German vernacular texts. He overtly rejects the authority of classical texts and thereby rejects the medical curriculum that others used to substantiate their authority as physicians. Physicians in other regions of Europe were able to assert their authority over the body through their knowledge of the inner workings based on dissection, but German doctors still tended to cite classical texts as the primary point of authority. Paracelsus rejects both approaches and argues that his approach to the practice of medicine recognizes the importance of the divine. His journey to understanding the sexed body shall be explored in the following chapters. For now it is important to recognize how Paracelsus's inquiry into human origins builds upon the fascination with human birth that was brewing in early sixteenth-century German regions. The increasing concern over satanic influence, and particularly the belief that women were more susceptible to Satan's luring, explains the popularity of the pamphlets on unusual births and the success of Rösslin's book warning against midwives. While on the one hand, popular authors relied on medical authority like Avicenna and Aristotle as a way to combat the fearful secrecy of the birthing chamber, on the other, authors relied on religious metaphors to assert the sanctity of human birth. Paracelsus was similarly interested in understanding what caused unusual births and his inquiry into the workings of the womb could be interpreted as an argument against women's culpability for unusual births. He, like many of his contemporaries, imagined the workings of the body and the imagination, because of concern for sorcery, was heavily occupied with thoughts about the working of the womb. Paracelsus's writing strikes a crossroads to vernacular tracts on birth, deformity, witchcraft and the arguments against untrained medical practitioners. He found himself caught between the established medical profession, which increasingly favoured university education or classical authority; and those who were outcast because of their religion or sex. Though his writing reveals in some ways a compassion for the otherness foisted upon these groups, he simultaneously seeks to establish his own medical authority and does so by rejecting the practice of Jewish physicians and the classical education associated with licensed physicians. Although he uses the topic of human birth as a way to carve his way into a respected position of authority, his contemplation of human birth ultimately leads him on a different path: toward a theory of embodied selfhood that includes a natural and spiritual body. Like other writers of his time, he also uses the female body, including its function in the process of conception and gestation, as a theoretical framework. Paracelsus's theory of embodiment emerges from his use of the female body as a rhetorical tool and then turns to comparative births of Christ, Mary, Adam and Eve to articulate a concept of spiritual and corporeal selfhood. In some ways the texts considered in this study have been misunderstood as poorly written expositions on a medical or theological theory; but I argue that we must read these documents as a writing process in which Paracelsus contemplated his own place in the medical profession and the meaning of a corporeal body.

2 PREDESTINED CONCEPTION: SEEDS OF PROCREATION AND THE WORKINGS OF THE WOMB

Paracelsus needed to invent himself. He decided that the best way to secure his prominence in the medical profession was to prove that he understood human birth better than other physicians who were profiting from providing advice on conception to wealthy clients. First, he had to explain the process of conception and gestation to himself and this is what he does in his earliest works. He uses his writing to tease out his understanding of human reproductive material and reproductive organs. He does not subscribe to the description of human reproduction found in classical texts that relied on understanding men as warm and women as cold or men as agents and women as receptacles for men's seed. Instead Paracelsus made a name for himself by rejecting the teachings of Aristotle and other prominent forbearers. If educated physicians were to explain the male seed as containing all of life, he would deconstruct that idea. If educated physicians were to disregard the importance of the womb he would emphasize it. He loved contradiction. He also had a strong sense of spirituality. This curious combination of rebellion and spirituality led Paracelsus to reflect on the meaning of the physical body and the degree to which God is involved in human births. This chapter explains the intersection of Paracelsus's quest for self-definition, his contemplation of the function of seeds, and his early emphasis on understanding the workings of the womb. The tracts under consideration in this chapter are considered his earliest writings, though the precise date on some of them is unknown. However, Paracelsus's quest to define his place in the medical profession as an expert on conception in these writings reveals a consistency among these tracts which led me to group them together. Because Paracelsus uses his writing to think through his understanding of the body, it is necessary to explore the tracts one at a time in order to focus on the process of his discoveries.

Book about the Birth of Rational Things Through Reason/Das Buch von der Gebärung der Empfindlichen Dingen der Vernunft (*c.* 1520)[1]

One of his earliest tracts, though the exact date remains unclear, *Das Buch von der Gebärung der Empfindlichen Dingen der Vernunft* (*c.* 1520) points to Paracelsus's need to differentiate himself from other medical practitioners by emphasizing his contrary understanding of conception.[2] He expresses his distrust of physicians who rely on the medical canon, largely based on the teachings of Aristotle. He asserts that reliance on Aristotelian medicine leads to errors in assumptions about human reproduction.[3] He believes that other physicians' inaccurate understanding of conception stems from their faith in classical texts.[4] Paracelsus emphasizes that his own faith rests on Christian teachings, on the Bible, and for this reason he declares that he is particularly concerned with how his contemporaries understood the materials and the process of conception.

He begins with what seems like a linguistic differentiation between his understanding of reproductive material and that of his contemporaries. Whereas he believes that seed (*samen)* is the essential reproductive element, he asserts that other doctors are misled by the idea that sperm (*sperma)* is the most important ingredient in human birth. Paracelsus dispels what appears to be a simple disagreement in terminology. He asserts that the difference between reproduction involving seed versus reproduction involving sperm represents the tension between medical reasoning that allows for God's action and medical reasoning that does not acknowledge divine influence. He believes that other physicians overlook God's involvement in human birth. With this stance, Paracelsus positions himself as distinctly different from the university-educated practitioners who believed in the truth of classical authority in defining sex-roles in reproduction.

Paracelsus needed to define his practice as an alternative to other non-university-educated practitioners, most likely Jewish physicians who claimed their authority by their expertise in herbals. To mark his approach to understanding conception as uniquely Christian, his model of human reproduction in *Buch von der Gebärung* rests on the idea that God has a special relationship to humans because Christ assumed human form.[5] He suggests that other physicians do not consider the similarity between Christ and humans and therefore disregard the closeness of humans to God. Not only do less-skilled physicians overlook the important connection between Christ and all humans, they sometimes are so foolish that they neglect to differentiate between humans and human-like beings. Suggesting that his contemporaries failed to distinguish between humans and human-like beings, Paracelsus explains that cyclopes among other natural beings, carry the materials necessary for reproduction within their bodies.[6] By contrast, reproductive material does not exist within the human body. Rather, humans have eternal spiritual qualities that other creatures lack. Because non-humans have no soul, seed (*samen)* is within these creatures.[7] On the one hand, he suggests that the lack of reproductive material in humans is a spatial concern because the soul takes the place of the seed.[8] On the other hand, Paracelsus regards the presence of seed (*samen)* as a human vulnerability that carries with it the potential for undesirable, even monstrous, births. However, his primary concern is to distinguish between the eternal (*ewigen)*, Christ and all humans, and the mortal (*etlichen)*, the creatures that have the materials for reproduction within them.[9] Grouping all humans with Christ substantiates Paracelsus's claim that his understanding of conception is far superior to his contemporaries who focus on sperm. He laments:

> Why do the blind doctors write so many chapters about sperm, as if sperm was the seed, and it is no seed. They want to use their art of concocting recipies to make women fertile, to produce sperm- this is beyond the capacity of their profession. For it is not a material process, it is a speculative (within the mind) process.[10]

Paracelsus's strategy to distance himself from other kinds of doctors by asserting the divine authority of conception leaves the impression that neither he nor the other doctors should provide aid in reproductive matters. He forcefully rejects the idea that doctors can increase fertility. Although he does nothing to carve a professional path for himself, he goes to great lengths to impede the lucrative practice of other physicians. In the process, he reflects on the relationship between all human birth and human's spirituality.

Paracelsus's insistence on the spiritual aspects of all human births leads him to question the relationship between the seeds of reproduction and human's mental activity; while he resolutely envisions the possibility that mental activity has a unique connection to God. In this way, his thoughts on conception lead him to articulate an idea of the physical body as distinctly different from the mental ties to spirituality. He asserts that reproduction 'is not material, but rather only mental' ('*ist nicht materialisch, alein speculuz*'), thus he emphasizes that the mental process provides the connection to God (as opposed to the idea that a divine power actually combines the reproductive material).

Because Paracelsus denies physicality in his rejection of conception as a material process, he must reconcile his understanding of divine involvement with the corporeal manifestation of conception. He is thus led to contemplate how humans receive the material for reproduction. He suggests that desire makes the seeds of conception (*samen*); but he also asserts that God places this seed into humans' '*Speculation*' (the mind). He explains conception as a mental process: "if the man wants, his mind makes him desire and the desire makes the seed, as the following will explain. So, God put the seed in the mind and gave the mind free will, to have desire or not." (*Wil der man, so macht ime sein speculation ein begirt, der begirt macht im den samen, als lauterer hernach folgen wird. Also hat got den samen gesetzt in die speculation und hat der speculation geben den freien willen, sich begirlich zu machen oder nit*).[11] There is a sense of confusion and mental wrestling in the writing as he seeks to understand mental and physical properties. For example, Paracelsus's use of the term *speculation* suggests both a physical location and mental activity. It is the place where God places the *samen;* because humans have free will that governs their *speculation,* they control their own thoughts/desires. Yet, he contradicts the idea that humans control their desires when he insists that God controls the act of desiring that leads to reproduction: 'God put the seed in the lust and the lust within the will of humans, what he wants shall be done' ('*also hat got in das verhengen gesezt den samen und das verhengen in den willen des menschen; was er wil das mag er tun*').[12] Although it is clear that Paracelsus questions the degree of divine involvement, it is evident that he thinks of mental activity as somehow distinct from physical properties. He explains, for example, that in order for reproduction to occur, God must simultaneously will a man and a woman to desire each other: 'And so when the lust is there, it must be in both, that is, in women and in men; therefore their mutual desire does the work of their will' ('*und so nu die verhengnus da ist, so muß sie in beiden sein, das ist in der frauen und im man; also ir beider verhengnus macht die werk aus irem willen*').[13] In this constellation the man and woman act as intermediaries in the divine plan. He is left questioning the extent to which humans have agency at all, the specific moments when divine influence is involved in human births, and the relationship between mind and body. All of these questions grow out of his desire to distinguish himself from other physicians.

Paracelsus's wish to explain human birth as a divine act leads him to question how godly influence could manifest itself in the human body. He explains this divine intervention as physically entering the thought process:

> God put the material of seeds in the will of humans that is taken from the object. So when he desires and his will fills his mind, then the liquid of life in the mind heats up and becomes the seed of another human. This is in men and also in women.[14]

He attempts to trace the process of conception as it occurs within the body, but at the same time Paracelsus must reconcile the notion that the seed is divinely willed with the limitations of the physical body. In other words, he imagines how God's will could physically enter the body and decides that the mind provides this portal. He explains that first God places the material of conception, the seed, in the human will; in other words, the seed is carried by desire or the wish to procreate. After this is done, and God's will is fulfilled through the human's desire, the *liquor vitae* is stimulated within the *speculation*, which refers to the mind. In this sense the workings of the mind have physical qualities, but Paracelsus also posits that a transformation occurs from the workings of the mind into the seeds of conception. The material moves from an imaginative state to a liquid then to some kind of solid when Paracelsus conjectures that the mind creates some kind of fluid from which the seed emerges. He calls this intermediary moment the *liquour vitae.*

Paracelsus has difficulty explaining the composition of the fluid and how it becomes solid, in part because his understanding of the process was comparable to his medical competition and he insisted that he had a profoundly different understanding of the process of conception. For example, at first the connection to *liquor vitae* is unclear to him as evident in the way he rethinks the concept. Paracelsus explains that the 'liquor vitae is none other than a hidden human' ('... *der liquor vitae ist nichts anders dan ein verborgener mensch*').[15] Furthermore, it is ' like a shadow on the wall, in the same way the seed is taken from humans and formed like him' (also, '*wie ein schatt an der want, derselbig nimpt sich von menschen und formirt sich nach ime*').[16] He imagines the *liquor vitae* to function like a reflection that resembles the human from which it is derived. This constellation of human conception explains the similarity of children to their parents. It also suggests how the human develops from divine material, and what the substance might be that God places within humans. However, the idea that the *liquor vitae* is a small human within the body is similar to his contemporaries' understanding of sperm. Always wishing to distinguish his theory from the idea of sperm, Paracelsus notes that the *liquor vitae* is an incomplete human, still lacking a soul, 'It does not have a soul. It takes its life only from the human body and not from itself...' '*wiewol er die sêl nicht hat und das leben von leib nimpt und nichts von im selbst*'.[17] Rather than explaining the process by which a human gains a soul, Paracelsus reminds his readers, and probably himself, that he intends to elucidate only the formation of the body.[18] He underscores his distinctive medical practice by distinguishing between the materials that form the body, that is the materials that are at some point within the human, and those that form the soul. He further distances his theory from the ideas of those he considers foolish doctors when he writes:

> Therefore understand what we say: you would like to believe that the seed of birth is found in the *vasa spermatica*, the way the foolish doctors talk about it, and that rubbing the *vasis spermatics* itself makes seed come out, the seed that they call sperm. That is the biggest lie that the doctors spread in order to make money and to prove their skills to people.[19]

He expresses his distain that such doctors use their theory for financial gain, and he rejects the straightforward physical description of how the material of reproduction is transferred. Paracelsus insists that his theory is more accurate and more complex than the theories of those who simply describe the process of reproduction in terms of the action and properties of sperm. He is convinced that reproduction is initiated by God at the moment when a man and a woman desire each other: only then does desire effect the needed physical change, namely when the *liquor vitae* is heated. Thus, he emphasizes that the seed from which humans grow does not enter into sperm: 'and the seed from which the human grows remains in the *liqoure* and it does not go into the sperm' ('*und der same von welchem der mensch wechst, derselbige bleibet im liqoure und kompt nicht in die sperma*').[20] He argues that doctors, who are eager to profit from offering help in conception, suggest that a human being is in the sperm. In Paracelsus's opinion such physicians are more interested in becoming rich than in uncovering the truth.[21]

By insisting that sperm is ultimately responsible for reproduction, he claims 'foolish' doctors have suggested that they could increase the chance of conception by selling elixirs to increase the production of sperm, i.e. potency. Paracelsus rejects the notion that sperm determines conception, arguing instead that both male and female contribute the materials necessary for conception and that human reproduction is part of a divine plan. He insists that no doctor can interfere with this process:

> The main point is that the *liquor vitae* is part of the body; thus the sperm retains this kind of form within it and has power to form itself like a liquor. But you should pay attention to the reason that the sperm cannot be born in a way that deviates from the form that is within the liquor...that is why the sperm is a part of the liquore vitae, which is a microcosm.
>
> *Dan ursach der liquor vitae ist ein form des leibs; also behelt der sperma ein solche form auch in ime und hat gewalt den selbigen auch zu formieren wie sein liquor. Aber ir sollet den grunt merken, das aus der sperma nichts anders geboren wird dan die form wie im liquore ist....darumb ist der sperma ein teil von dem liquore vitae, welcher ein microcosmus ist.*[22]

Paracelsus emphasizes that human reproduction, particularly the role of *liquor vitae* in it, must be understood as the microcosm of the universe, and the sperm is a further microcosm therein. However, sperm is incomplete. The necessary elements to form a complete human being are found in the seed (*samen*).[23] In Paracelsus's understanding of reproduction the *liquor vitae* incorporates both the idea of sperm and the idea of *samen*; this forces him to acknowledge that there might be some truth in the theory of reproduction that he seeks to discredit. To avoid such comparison, he puts forth a different understanding of reproduction that focuses on the mutual responsibility of men and women and divine involvement in conception.

In Paracelsus's theory of conception the process is initiated by desire, at which point God places the reproductive material in humans, after which desire causes the *liquor vitae* (which is in the mind) to be stimulated. Once the *liquor vitae* is activated, an image of the body is projected within the *liquor vitae* and a tracing of the image separates and becomes a seed. The man has seed and the woman has seed. In this paradigm men and women do the same thing in reproduction and their minds work in the same way to produce offspring. While one might think that Paracelsus was drawing on a Hippocratic construct of human conception in which both parents contribute seed to the offspring, Paracelsus envisions the process as alchemical: the seed is the result of a process of heating that transforms a material from one state to another. He also differentiates the corporeal workings of men and women. It takes him a bit of mental wrestling to come to this conclusion. At first he focuses only on the similarity of men and women's roles in creating the seed of human life. Because of his insistence on the duality of human reproductive material, Paracelsus must explain how these seed combine to form a child. While he views the mental processes of men and women as similar he doubts the physical similarities, which leads him to evaluate the importance of the womb in conception.

Unlike his contemporaries, he highlights the work done by the womb in conception. He asserts that seed is motionless and that the matrix, or womb, is needed to pull it to the necessary location. This understanding of the process of conception moves away from an understanding of the womb as passive receptacle, an idea most often associated with this time-period. In this way, he takes yet another stab at his contemporaries, but he is also led to rethink his theory. He begins to question whether or not man and woman have a complete seed, or if rather they each contribute half a seed to form something whole: 'The man has half a seed and the woman has half a seed and so the two together make a complete seed' ('*am ersten merken das der man ein halben samen hat und die frau ein halben; also die zwen machen ein ganzen samen*').[24] This change in thinking that postulates half a seed for each man and woman, rather than contributing separately a specific seed, calls into question his previous identification of seed (*samen)* as a product of the *liquor vitae.* He had earlier claimed that the *liquor vitae* contained all necessary elements for new life. As Paracelsus explores the power and efficacy of the matrix, his idea of *liquor vitae* becomes increasingly confused. His inability to accurately describe the *liquor vitae* focuses his attention still more on the womb. He writes,

> In the womb we find a pulling quality, which is like a precious stone and a magnet that pulls the seed towards it. Such a pulling power within the mother opens itself so that the seed can be born. At this point the work from the understanding that comes from the speculation is used and the womb pulls the seed of the humours to it from the woman and from the man, from hearts, from livers, from legs, from kidneys, from muscles, from blood, from skin and from the entire body. For whatever has a separate part in the body, it has its own seed.
>
> *In der matrix ist ein anziehende kraft, welche gleich ein agstein ist und einem magneten, an sich zu ziehen den samen. Solch an sich ziehende kraft in der muter kompt und*

> *eröfnet sich so der same geboren ist. Als dan so die werk nach dem verstant, der aus der speculation kompt, gebraucht werden, so zeucht die matrix den samen des humors an sich von der frauen und vom man, vom herzen, von [der] lebern, vom milz, vom gebein, vom mark, vom geeder, von musculis, vom blut, vom fleisch und von allem dem so im ganzen leib ist. Dan alles das so ein besonder stück ist im leib, das hat ein besondern samen.*[25]

This constellation of the womb reflects an alchemical understanding of the body as evident in his comparison of the pulling power of the womb and a magnet. Unlike his contemporaries who base their thinking on classical texts and assume that only the sperm has power to influence birth, Paracelsus highlights the agency of the womb: in this case the womb attracts and orders the reproductive material. Paracelsus attributes significant power to the matrix postulating that it has the ability to draw seed inward.[26] It also attracts sperm. However, since sperm plays an unimportant role in his theory of generation, Paracelsus theorizes that it is removed and discarded by the matrix. The matrix is responsible for arranging the seed; it is indispensable if reproduction is to occur. Paracelsus explains that when the matrix is unable to attract seed and discharge unwanted elements, no birth will result: 'when the pulling power of the womb is not there, there will be no birth' ('*und wo die anziehende kraft der matrices nit da ist, das wechset kein geburt. Dan an das ort kompt kein samen, wol ein sperma*').[27] He devotes the remainder of the tract to explaining how the matrix is responsible for pulling seed inward and arranging it in the correct way to form a healthy human. In this way, he overtly rejects the theories of university-educated physicians and instead puts forth an alchemical-based model of understanding reproduction that recognizes importance of the female body's work in conception.

His unconventional theory of conception and gestation allows him to redefine the widespread belief that women's thoughts influence the baby during gestation. Based on the idea that the womb pulls seed inward and organizes seed, Paracelsus contemplates the reason that some births are monstrous. He considers that this outcome results from the matrix attracting the wrong kind of *samen*, 'an unknown seed' ('*ein unbekannten samen*').[28] This reasoning puts him in an awkward place of rethinking his initial claim that monstrous births were not human births. He comes to terms with this dilemma by considering that non-human seed could enter the womb. However, he rephrases his explanation suggesting that animal *sperm*, not seed, is pulled into the matrix. He reminds his readers, and perhaps himself, that animals do not have seed (*samen)*: 'such a deformity is found only in sperm, for that which comes from animals is not a seed it is only sperm' ('*aber solches misgewechs sind alein in der sperma; wan was von tieren kompt das sind nur sperma. Wan im tier ist kein liquor vitae, alien ist ir sam in irer sperma*').[29] After he regains his footing by reverting to his initial argument that it is most important to understand human seed as something entirely different from sperm, he returns to his contemplation of the importance of the womb. He explains that when the matrix is unable to draw seed from a man or woman, it may draw in animal sperm instead. The offspring from such a conception will resemble animals.[30] Whereas Paracelsus suggests that other doctors believed that the primary material for reproduction was only provided by men, and that the womb served merely as a container, Paracelsus ponders women's role in conception further.

He is convinced that women contribute reproductive material, seed, but he also assumes that the womb plays an active role in conception. He suggests that the matrix puts the seed (*samen)* into the correct place. His idea of seed (*samen)* changes somewhat when he writes that each body part has its own seed. The matrix's job is to organize the seed and when all the seeds are correctly aligned to make a new human, the matrix's job is done. Though his notion of the active matrix contradicts his opponents' view of the male as the only active agent in conception, it does little to advance Paracelsus's distinction between his idea that all birth is divine and the divergent opinion that reproduction is within human control.

Only when Paracelsus describes the final stage of development does the connection to God through the womb become apparent. The growing child stays inside the mother until all the necessary vital organs are developed. At the very end of this process the offspring acquires soul, spirit and reason,

> so after that the spirit will be given, the soul, reason and everything that belongs to the soul. After that comes the moment of birth: when the soul and all its constituents do not remain in the mother's body, it hurries out and that is when the time comes
>
> so dan das also ist, als dan wird im geben der geist, die sêl, die vernunft der verstant und alles das so zu der sêl zugehorig sind. Als dan folgt hernach die zeit der geberung: wan die sêl und ir zugehörents bleibet nicht in muter leib, eilet heraus, wan die ziet ist da.[31]

Paracelsus concludes that the human gains a soul (*sêl*) while still in the matrix. Thus, from a theological standpoint, the matrix is important as the location where God intervenes to give humans a soul. Paracelsus's thinking about the moment when a human acquires a soul echoes his concept of the birth of Christ. In other words, he sees a parallel between the divine creation of Christ in Mary's womb and the creation of a soul within the womb of all other humans. Paracelsus links God to humans through the soul by way of the womb using a Christian schema of spirituality.

Paracelsus suggests that the woman's body plays a more significant role in reproduction than is customarily ascribed to her by the physicians Paracelsus regards as misguided. In his opinion, the so-called foolish doctors claim to increase the efficacy of sperm. Because these doctors assign the material of reproduction to males only, they disregard the significance of the female body in reproduction.[32] By contrast, Paracelsus downplays, to the point of ignoring, the significance of the male body in conception. His emphasis on the female, subsumed under matrix, was probably meant to contradict the beliefs of his contemporaries, rather than a conscious attempt to elevate the significance of women's bodies in reproduction. Either way, his approach to conception leads him to contemplate the similarity in the mental processes of men and women in the production of offspring, but he regards the corporeal qualities in different ways. In other words, he attributes agency to the womb, whereas his contemporaries distained such an approach. He begins to recognize unique and special qualities of the female body.

Paracelsus's recognition of a special connection between the mother and foetus is emphasized in several passages in *Book about Birth* (*Buch von der Gebärung).* First, he says that birth occurs when the offspring can no longer receive air or nourishment from the mother, 'the child grows in the union of soul and body until it is no longer nourished by maternal food and air. After that birth follows '*...so wechst das kint in die vereinigug sêl und leibs also lang, bis es sich des lufts nimmer enthalten mag und der müterlichen spies. Auf das folgt die geberung*'.[33] Next he suggests that what a woman eats while she is pregnant can affect the offspring. Although the mother would not turn into a crab when she eats one (because she is stronger than a developing child), the offspring may develop crab-like features because it digests some of the material through the mother.[34] He explains such a transformation further by saying, 'for all that the womb desires enters her and is not consumed in the stomach like other food' '*dan alles das so die matrix begert mit iren begirden, das selbig kompt in sie und wird nicht im magen gedeutet wie ein ander spies*'. The woman's stomach is unconnected to the foetus; rather, the matrix draws the material inward. Yet, elsewhere in the tract he says that a woman's blood could affect the child's development. In this, he contradicts a popular belief that a woman's fantasy could cause deformity in the child because it was believed that whatever the mother saw would be projected on to the baby. Here he dismisses the connection between mind and body that he previously outlined in his understanding of seed production in humans.

He emphasizes that the changes that occur after conception are a chemical process. Paracelsus dismisses the idea that a woman's thoughts could alter the contents of her womb by explaining that when the mother is frightened her blood flow will change, 'the blood that nourishes the child is drawn away from it or too much blood runs toward it and the nature of the mother in such a situation is altered. Through such a change the child falls ill' '*...das blut der narung des kints entzeucht oder zu vil hinzu laft und die nature der muter in solchem verenderd wird. Durch welche verenderung das kint in krankheit fallet...*'[35] His sentiments are protective of women in a way that would contradict the widespread belief that birth defects could reveal satanic influence on a mother. When he describes what today might be understood as metabolic changes that have the potential to alter the offspring he exculpates women from engaging in sorcery.

Whereas in the beginning of the tract Paracelsus overlooked the significance of the body in his attempt to explain God's control over human desire suggesting that male and female seed united mentally, in the latter portion of the tract, he suggests that the seed is somehow located within the body. He does not elaborate on the exact location of the seed; rather, he devotes the remainder of the tract to a discussion of the matrix, or womb. Again, although the womb represents a physical space, it remains somehow disconnected from the woman: It pulls seed in, organizes seed, and removes sperm. The foetus receives air and nourishment from the mother, but it cannot be influenced by a woman's thoughts. Though Paracelsus writes much about the womb and wonders about the relationship between the woman and the womb, overall he does not address corporeal issues in reproduction. At the predestined moment when a man and a women desire each other, the seed miraculously present themselves within the human, in the *Speculation*. How the seed ultimately combine to form a new human remains imprecise in Paracelsus's thinking, in part because he concerns himself primarily with divine intervention in reproduction. Owing to this preoccupation, he emphasizes throughout the tract his contemporary's lack of respect for the will of God. While he postulates that their conjectures about sperm reflect a misguided approach to understanding human birth, he stumbles when he tries to resolve the final stages of his seed theory. When he attempts to explain the significance of the womb in the reproductive process he loses sight of his argument in favour of the supremacy of God. His foremost concern may have been securing his own authority as a physician capable of assisting in conception. Since he already viewed himself as different from the university-based medical education that stressed the importance of studying the classical masters in understanding birth, he rejected those viewpoints using his understanding of Christ as a guide. His Christian approach to the body, which equated the divine birth of Christ to the birth of all humans, further aided him in distancing himself from Jewish physicians. Should a patient wish an alternative approach to the theories of Aristotle that emphasized the importance of the male seed, Paracelsus offered a different paradigm that emphasized a Christian understanding of the body that placed greater emphasis on the function of the female body than theories based on classical texts. Furthermore, his theory focuses on the spirituality of the patient and thereby suggests the importance of a Christian physician. Paracelsus's alternative, alchemical-based understanding of conception left him grappling with how to explain the workings of the womb. He returns to contemplate the womb and its importance in reproduction again in another tract, *De generatione hominis*.

De generatione hominis (The Generation of Humans)[36]

The exact date when Paracelsus wrote *de generatione hominis* is unknown. Karl Sudhoff estimates that it had to have been composed around the same time that Paracelsus wrote the *Buch von der Gebärung (Book about Birth).* The similarities between the texts are numerous. Both begin with the idea that humans are the most complete creation because Christ was human. In both texts, Paracelsus conjectures that man and women contribute something to the offspring. However, the *Buch von der Gebärung* and the *De generatione hominis* differ significantly in their understanding of seed.

In *De generatione hominis* it appears that Paracelsus still struggles to understand the production and function of seed. At the beginning of the tract he suggests that seed is located within humans, but that it lacks the potential to reproduce until God intervenes. Yet, he concludes the same passage by saying that humans have no seed at all.[37]

> God created human seed freely and without force, so that it lies still and hidden within humans, so that he neither knows of it, nor is burdened by it, nor forced to act by it. And understand human seed from which another human is born: that it is like a corn kernel in a field, it is as if it were dead and silent without any potential. The whole human is a seed. It functions in the same way. It is dead in not fruitful until a union occurs, like the field with the seed. So the birth of humans is nothing, there is within humans a birthing power: a singular human, woman or man, has no seed and has no seed within them.
>
> *Also frei und ledig one allen zwang hat got beschaffen den samen des menschen, das er also still und verborgen im menschen ligt, als das er das selbigen in keinerlei weis noch weg beschwert wird, noch gezwungen. Und verstehent des menschen samen also, von welchem das geboren wird das ander mensch: das er gleich ist wie ein kornsamen im acker, das selbig ist so vil als tot und stile, on alle wirkung. Also ist der ganz mensch ein samen und im samlicher wirkung tot und nicht fruchtbar, also lang bis ein zusamenfügung das geschiet, als der acker mit einem samen. Also sagen wir von geberung der menschen, das nichts sei, das dan sei im menschen ein geberende kraft: und ein ietlich mensch, es sei frau oder man, ist one samen und hat kein samen in ime.*[38]

Initially he suggests that seed (*samen)* is like a vegetable seed waiting to sprout. Then he shifts to the idea that humans themselves are seed only to conclude that neither man nor woman have seed within them. Here, he wrestles with the degree to which humans have generative potential without divine help. It is evident that he wishes to distinguish humans from other beings when he is cautious not to describe humans as too closely resembling animals and plants. For example, he explains that trees are born from their 'mother's' seed. He defines seed as an innate part of all living forms with the exception of rational creatures (humans), 'for in all growing things the seed is united with its mother, except in the rational things, they have no union with their mother' '*sonst in allen wachsenden dingen ist der samen vereinigt mit seiner muter alein in den empfintlichen dingen nicht, die selbigen haben kein vereinigung im samen mit irer mutter*'.[39] He maintains that humans are rational (*empfindliche)* creation, once again insisting on the connection between humans and God as he did in *Buch von der Gebärung*. The unique relationship between God and humans is further emphasized by the notion that God created the seed (*samen)* that makes humans.[40]

There emerge, however, some subtle differences between the ideas of human reproduction in *Buch von der Gebärung* and those in *De generatione hominis*. In both texts, Paracelsus highlights his antagonism towards doctors who suggest that sperm is the most important reproductive agent. However, in *De generatione hominis* he seems less concerned with what he sees as their misguided scientific reasoning, than with the idea that they are selling books for profit.[41] The two tracts also differ in the way Paracelsus understands his place in the medical community. In *De generatione hominis,* he frequently writes in the plural (wir) as if he had followers or as if he belonged to a group that shared his opinion. In contrast, Paracelsus here describes those who insist that sperm is the primary agent of reproduction as inexperienced physicians.[42] Referring to himself as an experienced physician, he explains that he recognizes that sperm is merely a cleansing agent and that seed is the most important element for new life.[43] This insistence on his professional experience differs from the way he comes across in the *Buch von der Gebärung*, where he emphasizes observation rather than his own expertise. Still, he portrays himself as a medical expert much more emphatically than he did in *Buch von der Gebärung*.

Though Paracelsus presents his role as a professional differently in *De generatione hominis,* in both tracts he emphasizes the involvement of men and women in reproduction through a kind of mental process. As rational beings, humans are given free will that, if exercised, prompts them to desire each other. This desire is born from their *speculation*.[44] Whereas in the *Buch von der Gebärung* he states that man and woman each had, or received, seed (*samen),* in *De generatione hominis* each draws on his or her half a fantasy:

> For God created the birth of humans so that it is in both humans and they are built for each other, the man and the woman for each other. Therefore each has a complete fantasy. For each has only a half a fantasy, for this is the way God created them.
>
> *Dan also hat got die geburt des menschen erschaffen, das sie stehet in beiden menschen und ist ie eine dem andern eingebildet, der man der frauen et econtra. Also wird in inen beiden ein volkomene phantasei. Dan ein mensch hat alein nur ein halbe phantasei, aber sie beide haben ein ganze phantasei, dan also hats got beschaffen.*[45]

Paracelsus emphasizes the opposing, yet unifying nature of the sexes, implying that God created humans as complementary opposites; the seed that results from desiring one another originates in God. It does not come from nature, 'You should understand from us that fantasy makes the seed and not nature' *'[Ihr sollt] von uns also in den weg verstehen und wissen, das die phantasei den samen macht und nicht die natur'*.[46] In both tracts, he strongly emphasizes that the substance for human reproduction does not spring forth from nature in the same way that a seed sprouts from the earth; rather, he postulates a divine connection between God and the development of new life. He is consistent in his support of the idea that humans are different from all other beings in nature, and that the seed originates with God; but he wavers on what might be the right moment when God assists in creating a child. In *Buch von der Gebärung*, he writes that God willed people to come together at a certain moment and he placed the seed in the speculation. In *De generatione hominis,* he maintains the important influence of predestination, but he does not mention any specific moment when a predestined couple simultaneously desires each other. Rather, he considers fantasy crucial for conception to occur.

When Paracelsus turns his attention to the womb, he focuses on the idea that it is the woman's destiny, rather than mutual (male/female) destiny to conceive. Furthermore, in *De generatione hominis* he is more certain about the womb's responsibility in conception and he acknowledges its importance. He describes the matrix as a locus where all the ingredients necessary for life combine. As in *Buch von der Gebärung*, here too, the matrix's ability to attract seed is especially important for reproduction to occur. When the magnet-like pull of the matrix fails to draw in human reproductive seed, no human being will form.

Whereas in *Buch von der Gebärung* Paracelsus insists that a woman's fantasy could not affect the child in the matrix, in *De generation hominis* he maintains that the woman's fantasy is necessary for reproduction. Although in *Buch von der Gebärung* he emphasized the detachment of the mother from the matrix, in *De generation hominis* he accepts the idea that her mind could influence the outcome of the conception. He explains that the matrix receives the seed after the woman's fantasy attracts it, but Paracelsus reminds his readers that the entire process is not willed by the couple but predestined by God.[47] Once in the womb, male and female seed duel. The offspring resembles the person whose seed is stronger, i.e., victorious. Even so Paracelsus claims that variance results from mental qualities: 'if the woman's fantasy is stronger than the man and heats the seed more, than it will make a girl' ('*so die frau in irer phantasei sterker gelidert ist denn der man und fester enzündet, der machet ein frauen*').[48] In this passage he does not separate mind from body as much as he did in *Buch von der Gebärung* (Book about Birth). By contrast, the imagination has the ability to alter what happens within the womb. In comparison to the *Buch von der Gebärung*, *De generation hominis* places greater emphasis on the woman's mental function required for conception, drawing on the popular idea that fantasy/imagination can affect the appearance of the offspring. He opposes such thinking in *Buch von der Gebärung.*

To complicate matters further, in a passage that reflects ideas more similar what are known to be his late works, Paracelsus contradicts the idea that God is involved in every aspect of human conception, even though he writes that gestation takes forty weeks, because God planned it that way.[49] He implies that nature works independently of the divine as he separates the action of nature from the divine plan when he discusses how the mixture of seed (*samen)* becomes a human. Here, he suggests that the first step of the reproductive process involves only nature,

> So nature stops and does nothing more to the child ... The main point is that it is a complete child and nature can do no more to it, for it has already influenced it. Thereafter the grace of God enters to complete the being, so that it comes to an end, it is all ordered and gives life.
>
> *Also hört iezt die natur auf und wirket nichts mer im kint...Dann ursach, es ist volkomen ein kint, und die natur kan weiter nichts daraus machen, dan sie hat ausgewirkt. Demnach so kompt die gnad gotes, nach dem und der gotheit zustehet, zu volenden ir geschöpf, das sie kompt an die ende, da sie hingeordnet ist und gibt im das leben.*[50]

Thus Paracelsus begins to distinguish between the power of God and the influence of nature. He rethinks his position that human seed differs from animal seed when he explains that nature's role in reproduction involves the initial composition of the body.[51] Contradicting his own insistence that God wills seed to come together, he asserts that nature produces and arranges the materials necessary to form a body and that God is involved only in the final stages of development. When the human shell, the human body, is complete, God makes the soul ('*das inwendige des Menschens*'). God is defined as the light that humans find '*liecht des menschen antrift*'.[52] This development contradicts his sentiments in *Buch von der Gebärung* and the beginning of *De generatione hominis.* One possible explanation for this disconnect might be that the last section of the tract was added at a later time. *De generatione hominis* first appeared in Latin well after Paracelsus's death in the autumn of 1575; the first German version appeared two years later in 1577.[53] Though Sudhoff noted the consistency of language in the two tracts, it remains unclear when Paracelsus completed the tract. The beginning of the tract may have been written around the time he wrote *Buch von der Gebärung,* and the conclusion may have been completed at a later date, or by a different author. These inconsistencies aside, the text reveals Paracelsus's conviction that human birth involves divine intervention, though he is unclear when and how this godly action occurs.[54]

The tracts *Buch von der Gebärung* (Book about Birth) and *De generatione Hominis* (On the Generation of Humans) give a glimpse into how Paracelsus's quest to define himself as a different kind of physician intersects with theological discussions of the day, namely when and how a human gains a soul. In the sixteenth century, reflections on the soul were intimately tied to understanding human birth, as the work of Kathleen Crowther has emphasized.[55] This correlation between a philosophy of the soul and a theory of human birth is further evident in the way modern theologians classify Reformation thinking about the soul. For example, George Huntston Williams, the foremost authority on the radical reformation, summarizes the three main views of when the soul joins the body in Reformation discussions of the topic: The soul is present '(1) from conception (Tertullian and traductianism); (2) at some stage in gestation; and (3) at the first breath "in the air of the world" (the basic Old Testament view in Gen. 2:7 and Stoics)'.[56] Paracelsus, never one to stick to a single path for long, considers all three of these possibilities at various stages in his writing. In this way he simultaneously responds to his medical competition and the concerns of theologians of his day.

The above tracts reveal his quest to explain the relationship between the seeds of human conception, the function of the womb, and the moment a body gains a soul. He ultimately determines that the embodied soul defines the human, but in these early tracts he wavers on how the soul emerges. He grapples with the idea of the seed containing the soul and the idea that the soul is something added to the process of conception at a later stage. Throughout his writing he returns to the question of when and how humans gain a soul, while at the same time he questions the meaning of human physicality. In his early tracts these questions emerge in the background of his quest to define his place in the field of medicine as the following tract further demonstrates.

Commentary on the Psalms[57]

Paracelsus's yearned for recognition as an expert on human reproduction and this desire resonated even in the works that compilers have labelled theological. For example, his 'Commentary on the Psalms', reviews ideas about conception that he explored in *Buch von der Gebärung* and in *De generatione hominis*. As in the other works discussed above, in his commentary on the psalms Paracelsus regards the womb as a divine place where God's will unfolds. The matrix is secretive and special because, Paracelsus explains, only God knows what happens in there.[58] Furthermore, Paracelsus's interpretation of Psalm 139 reiterates the idea that both parents contribute to conception, and that the entire process is willed by God.[59] In this context, he refers to the special relationship between God and the growing child. Only God knows the moment when the child is conceived, and, therefore, God knows the child before the parents recognize that they will become parents.[60] He explains that God does not want conception to occur at a specific, known time; that would make parents aware of the very moment of the creation of new life. Rather, conception should happen when it is least expected. He restates his opposition to those he called foolish doctors because they believed that sperm caused conception and that the process could be influenced by human intervention. Ignorance of the moment of conception made it impossible for physicians to manipulate the reproductive process. Similar to the previously discussed tracts, Paracelsus maintains his insistence on a kind of medical practice that recognizes the limitations of human power and the sanctity of the divine in human birth.

As he tries to sort out what makes people recognize the sanctity of God, he begins to articulate a sense that the physical body and the spiritual self function in different ways. For example, he writes,

> For what would the body recognize? Nothing! Is that not an enormous miraculous elevation in the eyes of God, who made us so miraculous, that only the soul functions in this way and not the body? For the body cannot understand, the soul however, not here, rather the smallest part here, but after our death we shall sense how powerful God is and how miraculous his work.
>
> *dann was will der leip do erkennen? Nix! ist das nit ein grosse wunderbarliche erhohung in got, der uns so wunderbarlich macht, das allein die seel wirkt und der leip nichs?...dan der leip mag nit begreifen, die seel aber; nit hie sonder das wenigst teil hie, nach unserm tod so wird sis empfinden, wie gross got sei und wie wunderbarlich seine werk seindt.*[61]

The sense of embodiment expressed in the above passage, that the physical and spiritual parts of humans function in different ways, is a concept that recurs in Paracelsus's works. It is not limited to the explicitly theological tracts like the one here, but rather develops both in relation to his medical interests and in regards to his understanding of spirituality. While an early sense of embodiment is evident in this tract, he maintains his insistence that he should be understood as an expert on human conception and this aspect of his writing carries over to one of his best known works: his tract on syphilis.

Von Ursprung und herkommen der franzosen sampt der Reception heilung acht Bücher (1529)/*On the Origins of the French Disease with Remedies for its Cure*

In the tracts examined thus far, Paracelsus addresses the topic of human conception as a divinely-willed process and avoids discussing coitus. However, his tract on syphilis (*Von Ursprung und herkommen der franzosen...*) he confronts the physical act of sex as a pathway for the spread of disease. He does not, however, agree with other physicians on the meaning of the sex act in the transmission of the illness, nor does he agree with their treatment methods. As in the tracts discussed above, Paracelsus is on a mission to define a new kind of medical practice and in his tract on syphilis he draws upon his theory of conception as a cornerstone of such practice. While Paracelsus's place in the history of medicine is often defined by his use of mercury to treat syphilis, it is important to recognize how even this treatment and diagnosis builds upon his theory of conception.[62]

He diverged from the opinion of other writers of the time who considered syphilis, or 'French Disease' (as it was commonly called), a contagious disease that revealed sinful or lustful behaviour. He also refuted the idea that it was an epidemic in the way that some medical writers following a Galenic model argued that it was spread through moist air or moist body parts, such as genitals.[63] Unlike his contemporaries, Paracelsus equates the spread of syphilis to the process of reproduction, not necessarily to the act of sex.[64]

His theory of human conception significantly influences the way he thinks about the disease and, most importantly, he wanted readers to make this connection. This expectation needs to be taken into consideration because this tract on syphilis, *Von Ursprung und herkommen der franzosen sampt der Reception heilung acht Bücher* (1529) appeared in print during his lifetime.[65] In this tract when Paracelsus begins to expound upon his theory of human reproduction, he refers his readers to 'die bücher de generationibus', presumably *De generatione hominis*.[66] He suggests that he has already achieved some kind of notoriety for his theory of human conception and that his theory of syphilis as a disease builds upon it. Thus, the tract on syphilis reinforces Paracelsus's desire to be known as an expert on human reproduction.

Paracelsus frames his understanding of the spread of syphilis in terms of human birth. In the same way that conception is a sort of chemical process involving what he sees as the smallest particle of life, the seed; the seed of syphilis operates in a similar way.[67] He explains that Venus gave birth to two daughters. The daughters spread syphilis. Although he describes the spread of disease through the metaphor of Venus, her children and their actions, he maintains the idea that both parents take part in creating new life. The importance of both parents in conception echoes his sentiments in his tracts on birth. He also returns to the idea that birth has divine origins.[68] He explains how the French disease hides in the body, lying dormant for a time before becoming visible: 'You know that in humans that the above mentioned illness originates in the body. It hides in humans for a long time before the visible outbreak occurs'.[69] This paradigm of disease echoes the way he wrote about conception in his commentary on the psalms. Furthermore, disease is hidden within the body, but is itself a body: 'if you recognize the nature of the body, you know the elements of illness, so you can heal them'.[70] In this sense, he views the disease as a microcosm of the body in the same way he regarded reproduction as a microcosm of the body in *De generatione hominis*.[71] His ideas of generation in his tract on syphilis are also similar to those expressed in *Buch von der Gebärung*, that is, God is ultimately responsible for reproduction, and both parents contribute reproductive material to the formation of the offspring.[72] However, in *Von ursprung und herkommen der franzosen* he occasionally uses the term '*sperma*' to designate the origin of life and disease.[73] This confusion in terminology arose in his other tracts on human birth too, but he is consistent in the idea that a seed with some kind of divine connection guides human birth:

> Because heaven controls coitus and its effect manifests itself on the sperm, like the effect that fire has on wood, so you should understand then that the child is dependent on the legacy (or heritage) that the father and mother make. The child may be like its mother or like its father. All of this is devised in heaven. The mother or the father could have illness and this would manifest itself in the seed. So the illness that began in heaven expresses itself on the sperm and it springs forth from there.[74]

Paracelsus signals his recognition that sexual intercourse is necessary for conception, and this too he sees as guided by a divine spirit. In the above passage he is most concerned with defining the divine origins of the seed. He asserts his conviction that all human birth, and he uses the same terminology that he did in *Buch von der Gebärung* (*empfindliches Geburt*), originates in heaven, continuing, 'for irrational begins do not come from heaven'.[75] Although much of the terminology echoes the ideas expressed in his other early tracts, Paracelsus goes a step further in his tract on syphilis when he directly deals with the act of coitus.

When he confronts the physicality of human reproduction, he immediately dissociates sexuality from the process of conception. This occurs when Paracelsus momentarily distinguishes between his theory of human reproduction and his theory of contagion. He writes: 'so Venus cooks the sperm during coitus and when overheated the illness lies within'.[76] Here, he acknowledges that disease can be transmitted through coitus. This reflection reveals an aspect that had been missing from his description of reproduction in *Buch von der Gebärung*: namely, the notion of physical exchange of reproductive material. Towards the end of the tract he does differentiate between the birth of disease and the birth of a human, between the birth of disease through coitus and human birth through seed (*samen*):

> Now recognise that it is a different spirit that makes seed to bring forth children. It is another seed that makes *luxum*. For *luxum*, or coitus, is not a generative force, rather it is an evil spirit that mixes with the virtuous generative seed and grows with it.
>
> *nun wissen hierin, es ist ein ander spiritus, der den samen macht, zu geberen die Kinder, es ist auch ein ander spiritus, der den luxum macht. Dan luxus oder coitus ist nit vis generativa, sonder ist ein spiritus peccans, der sich mit virtute generativa inbilded und neben ir mitlauft*.[77]

Although, once again, he recognizes that coitus spreads disease, he denies that human birth occurs the same way. Instead, he reverts to his theory of the divine seed suggesting that the materials unite through a mental process.

This tract's significance lies in the way it aids understanding of Paracelsus's theory of human birth. It demonstrates how he persists in his belief that there is a connection between the material of reproduction and God. It shows how he believed that his theory of human conception and the seeds of life were known to his readers. And lastly, the writing on syphilis reveals the extent to which he built upon his own unique understanding of the formation of human bodies, an event that seems to occur separate from coitus. While thinking about the origin of syphilis he implied an understanding of what today would be regarded as heredity, a recognition that the act of sex spread the disease, and a suspicion that somehow the traits of each parent had the potential to be passed on to the offspring.

In these early tracts Paracelsus carves out a reputation that relies heavily on his theory of the seeds of human reproduction. Paracelsus repeatedly suggests that human birth is part of a divine plan and it is a plan that he, more than any other physician, understands. The degree to which Paracelsus believes God to be directly involved in the reproductive process varies from tract to tract. In *Buch von der Gebärung, De generatione hominis* and his tract on syphilis, he conjectures that human seed has divine origins; in the commentary on the Psalms, he believes that God places the reproductive material in the womb and watches over the foetus as it develops in the womb. Paracelsus tends to disregard human physiology when discussing conception, but he continues to explore the importance of the womb in gestation. The question remains as to why he denies the corporeality of men and rejects the idea that reproduction involves coitus in many of these tracts. In his tract on syphilis, he emphasizes that the disease is spread through coitus; but he differentiates between the seed of illness and the seed of conception. He suggests that reproduction does not occur through corporeal contact, but rather through the creation of seed (*samen)* by a spirit (*spiritus)* which forms new life.

In the early writing three elements influence his thinking on reproduction. The first is that God is involved at some stage in the reproductive process; this idea is tied to the notion that humans are like Christ because Christ chose to be a two-footed, i.e. human, being. Secondly, Paracelsus dislikes doctors who are profiting from selling advice on reproduction, and he combines the idea that all human birth is part of a divine plan with his criticism of such doctors. Lastly, he explores the different notions about the origins of disease. In some instances, undesirable births are the result of the womb attracting the wrong kind of reproductive material; in other cases he suggests that unhealthy seed mixes with the reproductive material. Throughout the tracts reviewed in this chapter, Paracelsus suggests that both men and women contribute to reproduction, either with seeds (*samen,* or sometimes *sperma)* or through the combination of their fantasy. He implies that the womb has some connection to illness, but he does not explore the idea in detail. Thus far, the womb represents the only reference to the physical body in Paracelsus's theory of reproduction. While he claims to write about conception because he believes that many physicians are mistaken in their conviction that they can influence reproduction, he ultimately denies sexuality. Aside from considering the womb, he avoids discussing details of the physical body in his early works.

The influences on his approach to understanding the body were vast. While he engaged with contemporary theological discussions about the soul, he also incorporated alchemical-based thinking to human reproduction. While he rejected classical medical training based on the works of Aristotle, he never eschewed medieval medical ideas. His emphasis on the dual, and equal, responsibility of the parents in conception may stem from the medieval medical tradition. To medical writers of the fifteenth century men and women's bodies often did the same things and were seen as complementary opposites.[78] Caroline Walker Bynum has argued that this carried over to theological discussions and furthermore led to nascent definitions of a gendered body. She described how a theologian might interpret the blood of Christ shed in circumcision and on the cross as analogous to menstrual blood or milk. In Paracelsus we encounter the reverse. He employs theological concepts in describing human birth. He emphasizes that because Christ appeared in a human body that the human body must be closest to God. His emphasis on the divine origins and the holiness of human conception leads him to deny physicality and to focus on the miracle of conception. The sanctity that Paracelsus sees in conception leads him to depict physicians who provided (and profited from) reproductive advice as blasphemous. However, his concept of the holy body changes when he contemplates the real and tangent physicality of the womb. He is unable to reconcile this dilemma in the tracts considered in this chapter, but as the following chapters shall explain, he recognizes this stumbling block and seeks to explain it.

The tracts explored in this chapter present Paracelsus's earliest ruminations about the process of human conception and his repeated desire to find a way to envision God's involvement in all human birth. While his two-seed version of conception was not necessarily unique, Galen and Hippocrates insisted on such a model, his contemplation of the role of the womb in the process of aligning the seed suggests an understanding of the female body as distinctly different from the male body. This core principle in Pararacelsus's writing necessitates a reevaluation of Laqueur's assertion that, 'There was still in the sixteenth century, as there had been in classical antiquity, only one canonical body and that body was male'.[79] Laqueur drew on visual representations of the female in anatomical studies and argued that despite the fact that women may have understood the different nature of their bodies, the extant tracts available to us from the sixteenth century show how physicians tried to label the female body in reference to the male body. For example, the vagina was an inverted penis. The womb too was part of the inverted penis. The body was understood according to hierarchies, argued Laqueur. The male body was the pinnacle of this hierarchy. Paracelsus does not follow this patriarchal notion of hierarchy put forth by Laqueur. Instead, he privileges the female body in these early tracts when he emphasizes the importance of the womb in ordering and arranging the seeds in human conception.

Studying Paracelsus's approach to understanding the physical bodies of men and women provides evidence of theorizing the body, thus this approach to rethinking the claim of the origin of a two-sex model of the body differs from interpreting the findings of dissection, as has dominated early modern research to date. Although Paracelsus is known for his inconsistencies, he remains resolute in the idea that birth involves a divine element and the idea that men and women both play significant, and different, roles in the process of reproduction. This contemplation of his new kind of doctoring, and the roles of men and women in conception, ultimately leads Paracelsus along a path from which he does not stray. From his early thoughts on conception a theory of contagion and treatment of a sexed (male or female) body emerges. The following chapters will clarify how Paracelsus builds upon his early theory of reproduction to develop an understanding of the sexed body and the embodied self.

3 WHAT ABOUT MARY? CONTEMPLATING DIVINE AND HUMAN BIRTH

Paracelsus continues to pursue his reputation as an expert on conception and gestation in his theological works. The tone of his writing changes in these tracts and he becomes less accusatory and is instead reflective. It is evident that he begins to doubt his previous understanding of the relationship between God and humankind. In the tracts on conception (like *Buch von der Gebärung* and *De generatione Hominis*) Paracelsus's adhered to an idea of divine intervention in all human births and in those tracts he maintained that human bodies were like Christ's. However, in the tracts under consideration in this chapter when he attempts to explain Christ's birth as more special than any other human birth he is forced to rethink his previous theories. In this context, understanding Mary's role in the birth of Christ leaves him baffled. He wonders if her body was like other women's bodies, or if it is the nature of her body itself that makes Christ's birth special.

The tracts examined in this chapter are considered Paracelsus's theological writings, specifically his Mariological texts. Other scholars have tried to situate Paracelsus's approach to Marian-worship by aligning his ideas with Anabaptists or in some instances with more traditional reformers.[1] This approach illuminates the breadth of Marian theology in the German context, but neglects to connect Paracelsus's consideration of Mary with his medical interests. It makes sense that other studies would focus on Paracelsus's interest in Mary as a means to understand his theology. In the German realm one can trace the Marian theology back to the early fourteenth century with Brother Phillips 'Marienleben'.[2] And it is well-known that around 1500 fascination with Mary grew to the extent that vast collections of paintings and tracts emerged. At the point when Paracelsus writes about Mary, in the early part of the sixteenth century, theologians debated Mary's status: her virginity, her purity, if she had original sin, and if she ascended to heaven after death. Yet most reformers criticized what they viewed as an overemphasis on Mary.[3] While Paracelsus's writing on Mary may indeed demonstrate how he articulated his religious views, his fascination with her must also be understood in terms of his desire to define the human body, a body that he heretofore compared to Christ's.[4] In this sense, Paracelsus's interest in Mary's role as mother may be more similar to other reformers' contemplation of the Eucharist and the embodiment of Christ, than to the reformers' understandings of Mary because Paracelsus is particularly concerned with the degree to which Mary's body is manifestation of God's corporeality.[5]

Rather than tracing the similarities of Paracelsus's viewpoints to other Reformers, this chapter shall focus on the interplay between his so-called medical writing and his contemplation of human birth as it relates to the birth of Christ. While trying to understand the special role of Mary, Paracelsus begins to articulate the spiritual aspects of understanding Christ's body and he differentiates the body of Christ from the body of all other humans. I argue that this leads him to articulate his early theory of embodiment that understands humans to metaphorically contain a spiritual body, while the physical body is understood in context of the natural world.

My approach differs from previous scholarship on Paracelsus's Mariology. Katharine Biegger and Ute Gause are the foremost scholars on Paracelsus's Marian writing. Biegger contextualized Paracelsus's understanding of Mary in sixteenth-century theological discussions and demonstrated how Paracelsus's reputation was publically questioned when his Marian manuscripts were published in the latter half of the sixteenth century.[6] She emphasized the importance of Paracelsus's understanding of Genesis as the cornerstone to his theology and anthropology.[7] Biegger, like Goldammer before her, recognizes that Paracelsus's contemplation of the spiritual and corporeal aspects of humans resonated throughout his works, but this is not the focus of her inquiry.[8] Instead she explores the theology of his Marian writing in the context of other German Marian tracts. Ute Gause evaluates the numerous influences on Paracelsus's unique theology concluding that his thoughts were an amalgamation of contemporary and medieval theology and natural philosophy. Gause recognized how Paracelsus's thoughts about Mary led him to conceptualize human spirituality.[9] I agree with Gause's analysis, but focus on the interplay of Paracelsus's medical tracts in helping him to develop this concept.

Other scholars who focus on Paracelsus's theological works spend less time on his contemplation of Mary, but assess his theological interest more broadly. Many of these studies attempt to situate Paracelsus within contemporary theological discussions and to find sources for Paracelsus's theological ideas.[10] The conclusions reached by most scholars in the field point to the difficulty of situating Paracelsus in any one particular theological movement. While we may never know which theologians most influenced Paracelsus, he expressed his desire to be known and respected as a physician; therefore this study focuses on the relationship between his self-definition in the field of medicine and his theological interests. In the tracts considered in this chapter, he continues to attempt to define the process and significance of human birth by contrasting it with the birth of Christ. The ruminations about Mary's role in Christ's birth lead Paracelsus to rethink his understanding of human births, which in turn leads him to delineate the corporeal and spiritual qualities of humans.

Liber de Sancta Trinitate (1524)[11]

The tract *Liber de Sancta Trinitate* explores the formation of the Trinity, human mortality and the connections between humans and God. His attempt to explain the trinity leads him to clarify his thoughts about the nature of the human body.[12] His introduction to the tract, which is dated, declares: 'We want to talk about beliefs here, not about the body' (*hie wollen wir aber vom glauben reden und nit vom leib*).[13] But, as is often the case with Paracelsus's writing, he does not follow through with the original aim of the tract; instead, he actually does reflect on humans' physical body.

Liber de Sancta Trinitate is significant in the development of Paracelsus's understanding of embodiment because he begins to articulate an understanding of the dual nature of humans. In addition, he articulates a core principle of his chemical theory, namely separation. Bruce Moran explained how Parcelsus's concept of the origins of the earth through separation from God created the three primary elements. These elements: Mercury, Sulphur and Salt are central to all of Paracelsus's chemical philosophy.[14] Although Moran acknowledges the important role reflection on the divine played in Alchemical thought of the medieval and early modern period, he does not delve into Paracelsus's theological writing or explain how Paracelsus's contemplation of the divine influenced his chemical theories. Yet Moran rightly explains how the theory of separation helped Paracelsus explain the inner workings of the body.[15]

It is important to understand the ties between Paracelsus's so-called theological tracts and his chemical writing. The idea of separation is apparent in several theological tracts. Indeed, Paracelsus explores the initial separation of God himself and then separation of Mary from God. Whether this is an alchemical principle informing theological interpretation or the other way around, we may never know. However, when considered in the context of Paracelsus's continued reflection on human birth, *Liber de Sancta Trinitate* reveals how understanding Mary leads Paracelsus to refine his thoughts on the natural and spiritual bodies within humans. He begins the tract by explaining that originally God had one body until he decided to will himself into three entities:

> He was physically without a beginning, with blood and flesh that was tangible and visible to us and so it is still and remains in all eternity. Based on this he created his likeness the same way with mortal blood and mortal flesh. Not therefore exactly the same, for the human is not the same as God because of the soul. But he is like God

> because God had blood and skin. So there is not much difference between God and humans, because humans are like God because of the soul.
>
> *Er ist gewesen leiblich ohn*[16] *anfang, mit plut und fleisch greiflich und sichtbarlich und ist noch und bleibts in ewigkeit. Uf das hat er beschaffen ihme gleich das tödlich blut und das tödlich fleisch. Nit also, daß der mensch gott gleich sei von wegen der seel. Darumb ist aber er ihm gleich, in dem daß gott auch blut und fleisch hat. Also auch der mensch, und ist zwischen gott und dem menschen keine ander undterscheid, dann daß der mensch gott gleich sei von wegen der seel.*[17]

Focusing on the body, Paracelsus constructs an analogy between the body of God and the human body. Gause contends that Paracelsus recognizes the bodily characteristics attributed to God in this context as spiritual attributes.[18] She assumes that Paracelsus's thoughts on the eternal nature of God distinguish humans physically as mortal. Indeed, Paracelsus writes, 'God is the only god, the eternal god over all humans, for his blood and flesh, his form, his body remains in eternity, but humans will be eaten by worms and will rot. And nothing goes from him to God, only the soul' (*'allein sovil ist gott, der einig gott, der ewig gott, über den menschen, daß sein blut und sein fleisch, sein formb, sein gestalt bleibt in ewigkeit, und den menschen fressen würm, die kroten. Und nichts kombt von ihm zu gott, allein die seel'*).[19] While recognizing that Paracelsus does begin to emphasize spiritual qualities of humans and the eternal nature of God in *Liber de Sancta Trinitate*, as Gause postulates, the formulation of Paracelsus's thoughts on the divine in terms of corporeality requires further investigation. As the above passage demonstrates he begins to distance aspects of human embodiment from the divine by positing that the physical body will decay and remain in the natural world. Paracelsus finds it necessary to delineate the spiritual from the corporeal aspects of humans and he does this by reflecting on the significance of Mary.

In the texts discussed in the previous chapter Paracelsus argued for the similarity of humans to Christ based on physical attributes. Here, he emphasizes the prelapsarian nature of God as a concrete and thus visible body. Humans resemble the original body of God. Paracelsus's differentiation between humans and God on a spiritual level emerges with his concept of spiritual life. He explains the spiritual qualities of humans as a second light: 'There are two lights: human and spiritual and both come from God, namely the light of knowledge and the light of human life and the light of spiritual life' ('*Zwei liecht seindt, menschlich und geistlich und komben beide von gott, nemblich das liecht der weisheit und das liecht des menschlichen lebens und das licht des geistlichen leben*').[20] Still, he clings to the idea that the human form is most closely related to God. In an attempt to sort through his thoughts about the similarity of humankind to God, he turns his attention to the significance of the body in relationship to the Trinity. Considering the difference in the nature of the Trinity and that of humans, he explains the special, unfathomable nature of Mary.[21] While she lived in a woman's body, she was not separate from God, 'For she is and remains a woman in bodily form, but in a spiritual sense she is complete and just like God the father, for she retained the godliness from which she grew'. *('...dann sie bleibt ein frau in der person, und ir bleibt fräulicher gewalt. Aber in der gottheit ist sie volkomben und gott dem vatter gleichförmig. Dann sie seindt eines bleibens in der gottheit, daraus sie komt')*.[22] Paracelsus explains that Mary emerged from God in a parallel way that Eve grew from Adam. He carries the analogy through to the gendered expectations of Adam as powerful and Eve as weak. He further explains that blessed women, like Mary, have no power in and of themselves, but the Holy Ghost has authority to influence or change all things.[23] To Paracelsus's thinking Mary is transient. Andrew Weeks recognized this unusual understanding of Mary in Paracelsus's *Liber de Sancta Trinitate* as both a person and not a person. He rightly points out that the tract reflects Paracelsus's early ruminations about dualistic properties. Weeks refers specifically to the theory of two lights of nature.[24] However, the physicality of Mary piques Paracelsus's attention in another way. He reflects on whether or not her whole being is divine, or simply her womb.[25]

As in previous tracts, Paracelsus struggles to posit the womb as an influential reproductive organ within women. In *Liber de Sancta Trinitate*, he explains that Mary is the body of God in female form, existing briefly for the sole purpose of giving birth to Christ.[26] Paracelsus tends to diverge from Reformation thought when he distinguishes between Mary as a spiritual entity and Mary as a body. Since her exceptional nature lies in her body, Paracelsus explores her physicality relative to her role in the birth of Christ. He explains Christ's conception as a union: 'of two people, namely God and the Goddess, the holy spirit from God the father ... from both in one Godliness' ('*von zweier personen, nemblich von gott und der göttin, der heilige geist von gott dem vatter und von gott dem sun, aus beiden personen in einer gottheit*').[27] He does not say that Mary contributed any substance to the formation of Christ, but that, nevertheless, she was integral to the birthing process. Furthermore, he elevates her to a status equal to God, emphasizing the corporeal relationship between them. Mary was directly of God's body and thus eternal:

> The queen is made from the body of god, pulled from flesh and blood into an eternal flesh and blood, an eternal woman, unbreakable, virginal and unmarried, and for the main purpose that she is part of the trinity, it is her husband or her son or both their spirits.
>
> *die künigin aber ist von gott aus seiner person gemacht, gezogen im blut und fleisch und eines ewigen bluts und eines ewigen fleisch, eine ewige frau, unzerbrechlich, unbefleckt und unvermähligt, aus ursach, daß ein person in der trinitet ist, es sei ir mann oder ir sun oder ir beeder geist.*'[28]

The insistence that Mary must be divine harkens back to his understanding of conception in *Buch von der Gebärung* in which both parents play equally important roles in contributing seed to form the offspring. In this way, the eternal and divine natures of Mary facilitate the construction of a paradigm of Christ's birth that distinguishes it from the birth of other humans: He emerges from the union of two divine bodies. Paracelsus's confusion about Mary's corporeality emerges when he suggests that Mary does not exert any influence on the birth of the son, but rather acts as an intermediary for God and the Holy Spirit.[29] She herself lacks all human attributes. Leaving aside further exploration of the way Christ was born from Mary, or Mary's role in the Trinity, Paracelsus shifts his attention toward humankind. He decides that humans are like God and refers his readers to his other works: 'Because we are eternal, without end and we do not die, it follows then that we are like God. You will read more about this in the *paramirischen* tracts'. '*aber von deswegen daß wir ewig seindt, ohn end und nicht sterben, folgt aus dem, daß wir gott gleich sent. Darvon ir in den paramirischen schriften von ursachen der beschaffung weiter lesen werden*'.[30] Unable to explain the body of Mary, Paracelsus reminds his readers of the similarities of humans to the divine and emphasizes his expertise on the topic by referring to his other writing. He further explores the significance of Mary and whether or not her body should be considered human in his tract *De Invocatione Beatae Mariae Virginis*.

De Invocatione Beatae Mariae Virginis (*c.* 1527)[31]

In *De Invocatione Beatae Mariae Virginis* Paracelsus distinguishes between holy bodies and earthly bodies. However he has trouble fitting Christ and, indeed humans, into these categories. He postulates that there are two kinds of beings, *die creatur* and *die heiligen,* and he explains that only *die heiligen* are directly created by God.[32] As a *creatur* Mary differs from Eve. Christ was born from an exceptional female, rather than from a normal woman:

> So it is that we should understand Mary. She is a *creatur*. And if one reads scripture the right way, than she is the first *creatur*, before heaven and earth were created, thereafter heaven and earth were created. Next came the human. Now recognize the difference between them. The *creatur* is created from God, and the *creatur* from which he created humans from the earth. How different they are from one another, the heaven and the earth. For this reason there are two *creaturs*: one from earth and one that is not. The other from the Holy Ghost is alive. Mary is from that very same spirit. And Christ is no *creatur*, but rather he is born from God in the form of a *creatur*. So you see what advantages Mary had over Eve and her daughters.
>
> *Also ist nun von Maria auch zu verstehn. sie ist ein creatur. unnd so die geschrifft recht geleßen wirdt, so ist sie die ersste creatur gewesen, ehe das himel unnd erdten beschaffen sein worden, darnach seindt himmel unnd erdten beschaffen. darnach der mensch. Nun*

> *besich die unterscheidt von einander. die creatur die Gott im selbs geschaffen hat, unnd die creatur, die er den menschen von der erden beschaffen hat. wie weit es von einander stiendt, das himblisch, das irrdisch. darumb seindt zwo creatur: die von der erdten, die ist nichtz werdt. die ander von dem heilligen geist, die ist lebentig. von derselbigen ist Maria. und Christus ist kein creatur. vom vatter aber geboren in ein gestallt der creatur. So siehe nun auch weitter, was vortheil maria hat vor Eua unnd ir töchter.*[33]

Mary originated before Creation as a special human-like being. However, unlike other *creatur(en)*, God created her. This passage suggests a shifting away from the ideas expressed in earlier tracts, namely that humans maintain a special connection to God because Christ chose to be a two-footed creature. Here, Paracelsus states that Christ and humankind resemble each other only in form, in body. By positioning Christ closer to God than humankind, Paracelsus differentiates between the birth of Christ and that of other human births because of Mary's distinctive status. He moves away from explaining the exceptional nature of Christ through his conception, and he defines the uniquely divine nature of his birth as related to his mother's special status.

Paracelsus's insistence on the importance of Mary's special nature in the birth of Christ reflects his thoughts on conception in his so-called medical writing. Namely, he regards the woman's body as integral in conception and gestation. Further drawing on his construction of women's role in reproduction as postulated in the texts encountered in the previous chapter, he argues that descendants of Eve are tainted, as are all women. Christ could not have been borne of women like Eve, 'not the fruit of someone who was a whore and would taint her son with sins and other vices' ('*nitt die frücht einer, die vohler hurereu steckhet unnd ir söne vohler laster unnd sündten*').[34] Unlike the rest of humanity, Mary is not related to Eve, for descendants of Eve lack a comparable connection to God.[35] Mary's important role in Christ's birth reflects Paracelsus's conviction that mothers influence children during gestation and participate in creating offspring. Furthermore, as in the tracts discussed previously, he continues to attempt to define the nature of the womb.

Mary's status as embodying all of creation in her womb leads Paracelsus to suggest why she should be worshipped along with the trinity. Her body is described as all-encompassing, almost like the macrocosm. In other words, he envisions Mary as he describes the entire natural realm which leaves the impression that he seeks to distance her body from all other women. In an unusual passage in the tract, Paracelsus imagines Mary saying, 'God, who created me and all of you, heaven and earth, these I carry under my breasts' ('*Gott, der mich unnd euch alle, auch himmel unnnd erdten beschaffen, den trag ich undter meinem bußen*').[36] Even though he describes Mary's physicality in a mystical way, his explanation of Christ's conception are similar to his thoughts on human conception.

The language Paracelsus uses to describe the conception and gestation of Christ echoes his explanation of the human reproductive process in several of his other early medical works (as described in the previous chapter). His understanding of the material of reproduction in *De Invocatione* is consistent. In a fragment of the text he explains that the seed *(samen)* from Mary (that is directly created by God) remained untainted by human nature because, Paracelsus explains, God said that Mary was born: ' only from my seed. That is, from the fruit that I have carried, you shall all be protected' ('... *allein von meinem samen Das ist von der frucht, die ich getragen hab, werdent ir inn dem allen gewahrt*').[37] Explaining that by fruit *(frucht)* he means seed *(samen)*, Paracelsus confirms the connection that must be drawn between his theological and his medical writings.[38] It should also be noted that the way he uses the term seed *(samen)* in this tract generally agrees with his use of the term in the *Buch von der Gebärung* and *De Generatione Hominis*. Although, those tracts compare all human births ('two-footed creatures') to Christ, the *Invocatione Maria Virginis* distances Christ from any association with the human female. He insists that Mary must not be mistaken for a normal human.[39] In the 'Magnificat', which Biegger believes to be part of *De invocatione Maria Virginis,* Paracelsus moves the birth of Christ even further away from a woman's body and complicates his corporeal understanding of the Son: 'For she bore Christ in her soul and in this way intensified God's glory, for nothing corporeal happened here, no conception, no birthing the way a pregnant woman would. She remained bodily untainted' '*Nun hat sie in ir seel Christum geborn, damit hat sie Gott groß gemacht und in großem lob behalten, das nichts laiblichs dahin zu ist komen und empfängknus, geburt unnd schwangere art. als in der seel beschechen, on laibliche befleckhun*'.[40] Unlike other women, Mary's soul took on corporeal qualities and acted as a womb for Christ. It appears that his theological writings underscore his longstanding contemplation of the dual qualities of humans: visible and invisible, or physical and spiritual. Thinking about Mary leads him to assume that the soul as an invisible body could provide the space for gestation. In this way the spiritual body within all humans manifests itself in a corporeal way in the divine figure Mary. Paracelsus differentiates here between ordinary humans and special humans, like Mary. He knows that all humans are related through the seed *(samen)* that was passed on through Adam and Eve, but that Christ was the chosen son of God and, therefore, he did not grow in a human womb. He suggests, therefore, that to be human requires gestation in a human womb. This idea strongly contradicts the myth that Paracelsus came up with a remedy to make human life in a laboratory. Based on the tracts that have been considered thus far, he attempted to make a name for himself by recognizing the sanctity of human life and his contemplation of the special role of Mary in Christ's birth underscores his core belief that the womb plays a significant role in determining human life. Thinking about birth and Mary also demonstrates how Paracelsus begins to understand corporeality in a way that distinguishes it from spirituality. He continues to consider human corporeality and human spiritual identities in *De Genealogia Christi*.

De Genealogia Christi[41]

Like many of the tracts already considered in this book, the dating of *De Genealogia Christi* remains imprecise. Based on the terminology that Paracelsus employs in *De Genealogia Christi*, Kurt Goldammer hypothesizes that the tract was written either around 1530, or a bit later between 1530–7.[42] Goldammer was convinced that this tract is one of Paracelsus's late works. However, there are some passages that coincide with ideas expressed in what Sudhoff suggested was Paracelsus's earlier writing. For example, Paracelsus classifies humans as 'sensitive' '*empfindliche*[n]' creation describing the reproductive process as a shadow projected on the wall.[43] Such references recall the way he explains conception and the *liquor vitae* in the previously discussed tracts, particularly *Buch von der Gebärung* and *De generatione hominis*. However, the similarity in language aside, Paracelsus makes some philosophical advancement in understanding human embodiment in this tract.

In *De Geneologia Christi* Paracelsus expresses ideas about the spiritual body that he was unable to clarify in the tracts discussed thus far. Since most Paracelsus scholars are in agreement that all the texts I have addressed up to this point are indeed early works, I wish to suggest that *De geneologi Christi* demonstrates a transition in Paracelsus's thinking and should be considered the beginning of his more mature writing.

De Geneologia Christi differs from the other early works (examined in Chapter 2 and those above) because in this tract Paracelsus explicitly focuses his attention the birth of Christ using it as a metaphor of human spiritual birth (when Christians recognize Christ as a saviour). He articulates in this tract a concept of spiritual birth in a way that suggests that the ability to believe is itself divinely willed. In this sense, Paracelsus is beginning to contemplate a spiritual birth, as opposed to the strictly physical birth he focused on in other writing. He explains that those whose birth God willed embody the trinity and receive eternal life.[44] Later, he explains the birth of eternal qualities as part of the predestined nature of all Christians. That is, the corporeal body contributes to the ability to believe in Christ and gain salvation through that belief. Paracelsus explains that such a birth is predestined through God's planned marriages: 'Marriage is for the sake of children. For this reason God wills people together, to have children as He wishes' ('*Dann von der kinder wegen soll die ehe sein. darumb soll sie gott zusamben füegen; der weißt, wie er kinder haben will nach seinem willen, und nit nach willen des menschen*').[45] The union of parents implies that Paracelsus still considers divine involvement in conception, as he did in the works considered previously. Although this schema of conception differs from his understanding in *Buch von der Gebärung* because willing man and woman together differs from directly placing the material of conception within their bodies. Yet Paracelsus does not abandon the idea that some divine material influences the child's future.

When Paracelsus writes of destiny and the birth of children he suggests that corporeality is involved in reproduction through the union of the parents. He equates this predestined birth with faith. He maintains that all Christians are born in this way and that non-believers are the result of an evil influence on the union of two people whom God did not intend to join.[46] Building on the idea of destiny, Paracelsus turns to innate human qualities. He argues that, because belief in Christ begins before birth, forced baptism is ineffectual because humans cannot change what is predestined to occur.[47] He suggests that the ability to believe as a Christian is an innate quality over which humans have no control. In this sense, he regards faith as a physical, invisible quality that is only explained through understanding the moment of conception. Furthermore, he articulates a sense of heredity: being a Christian is not a process of belief; but rather like a seed that that exists within the human body.[48]

Paracelsus uses his contemplation of the birth of Christ as a philosophical exercise to understand the qualities of human bodies and human spirituality. In the opening of the tract, Paracelsus explicitly refers to the corporeality of Christ in relationship to the Trinity.[49] He insists that corporeality is tied to God. He explains that he wishes to make clear that birth of the human body relates to the birth of Christ.[50] Although he insists that he writes of the body as it relates to Christ, he conflates spirituality with corporeality. He suggests that understanding the supernatural relates to understanding the natural things among which he counts humans.[51] In a sort of circular fashion he explains that nature is like the trinity, which was created by God.[52] He clearly recognizes a relationship between God and the natural world, but struggles to define exactly how they function together.

Even though Paracelsus wavers on his definition of the natural world and the relationship of the trinity to natural order, he begins with the consideration of Christ's body which in turn leads him to clarify his thoughts on the human body. He suggests that after the birth of Christ, there emerged a new kind of human, one with eternal qualities.[53] He clarifies his understanding of faith and corporeality by explaining death, 'As I have said, everything that is without a beginning is also without an end. Where there is no beginning there can be no birth. What is never born cannot die. So it remains in the trinity without beginning and without end' (*'Nun hab ich gemelt, daß alles, das ohn ein anfang sei, das sei auch ohn ein end. Und wo kein anfang ist, da ist auch kein gebern gewesen.*

Was kein geburt ist, das ist auch kein tod. Also bleibt nun die trinitet ohn anfang und ohn end').[54] In his evaluation of death, he mixes physiological and theological ideas. Whereas he wrote of the predestined birth of those who embody the Trinity, he must reconsider the limitations placed on corporeal birth when he asserts that in death everything returns to its origins.[55] He explains that those born from God are eternal and that those born 'from the earth from flesh' ('*aus der erden vom fleish*') are mortal.[56] This distinction between eternal and mortal clarifies Paracelsus's theory of embodiment because he begins to delineate the physical and spiritual births of humans as distinctly different. For example, in *De genealogia Christi*, he considers that humans are created in two ways,

> In the same way that we come from earth and must return to the earth, we should know that a human is twice created: first from the earth and then the other way, from God's inspiration.
>
> *So wir nun auf erden komben seindt und wiederumb in die erden müssen-dann wir seindt aus derselbigen-, so sollen wir zum ersten wissen, daß ein iedlicher mensch ist zeifach geschaffen, nämblich und zum ersten aus der erde, zum anderen aus der inspriation gottes.*[57]

This double creation implies a corporeal and a spiritual birth, though Paracelsus's formulation remains opaque. He implies that humans have earthly and spiritual ties, but he fails to define the earthly limitations. In considering the human body in comparison to Christ's he elaborates his understanding of the eternal and mortal bodies:

> So notice further, that no body goes from the earth, that of Adam, that of human seed, that is why his body with his corporeality is not dead, rather it remains in eternity. Corporeality, if one is mortal and sensitive, is without a body from the earth, as I have explained the human body, rather it goes from Christ in its death, the earthly is different from him. It is like a shadow of a man on a wall, and that same man does not die. When he is buried, his shadow is buried with him. Therefore the body of Christ hung on the cross and died. The mortal body did not leave him until the third day, when he rose again. So, the body is nothing other than shadow, for immortal bodies make no shadows.
>
> *So merkent weiter, daß gar kein leib von der erden geet...noch von Adam noch vom menschlichen samen, darumb ist sein leib mit seiner leiblichkeit nit gestorben, sonder bleibt in ewigkeit. Die leiblichheit, so er tödlich und empfindlich an ihm gehabt hat, ist ohn ein leib gewesen von der erden, wie ich gemelt hab von des menschen leib, sonder sie ist von Christo gegangen in sem tod, da von ihm geschieden ist das irdisch gleich als ein schatten, der von eim mann gehet an ein wand, und derselbig mann der stirbt. So er nun vergraben wird, so ist sein schatten auch vergraben. Also der leib Christi am kreuz gehangen und am kreuz gestorben. Der tödlich leib ist aber nit von ihm komben so lang bis an den dritten tag, da er erstanden ist. da ist der leib nichts als ein schatten gewest, dann die untödlich leib geben kein schatten.*[58]

The difficulty of this passage notwithstanding, Paracelsus works through the notion that Christ had a human body. Recalling that he asserted in *Buch von der Gebärung* that Christ's body was human, here he considers that it only appeared human, much like his formulation in the other Marian tracts. He explains that, unlike humans, Christ's body lacks a materiality.[59] By contrast, he maintains that humans have both eternal and mortal qualities within them, 'in our bodies, death and eternity are side by side' ('*...in unser fleisch, tödlichs und ewigs beeinander*').[60] Even though he went to great lengths to distinguish Christ's body on the cross from a human body, he contradicts the distinction he drew between humans and Christ when he writes that the death of humans reflects the death of Christ. Both the immortal and mortal bodies will ascend to heaven after death.[61]

Although Paracelsus tends to read the death of Christ as a model for human experience, he suggests in *De Geneologia Christi* the possibility of a second birth, a birth that is a metaphor for receiving faith:

> So notice first, that we have a calling and a holy writing that guides us, that we need to be born again, namely not from human seed nor from the belly of women, but rather from above through the water and the holy spirit.
>
> *So merkent anfänglich, daß wir haben ein euangelium und ein lautere geschrift, die da uns ausweist, daß wir müessen zum andern mal geboren werden, nemblich nit vom menschensamen nocht wiederumb von dem bauch der frauen komben, sonder von oben herab durch das wasser und den heligen geist.*[62]

In *De Genealogia Christi*, however, Paracelsus suggests that predestined birth facilitates belief in Christ and the Holy Trinity.[63] The above passage indicates, however, that he regards physical and spiritual birth as distinct. The corporeal birth requires a human womb, the spiritual birth requires a divine connection.

While he makes significant strides in articulating how the human has dualistic properties: corporeal and spiritual; he stumbles in defining the physical qualities of Mary. He begins by explaining that God conferred special knowledge on Mary, since she was different from other humans: 'Adam was created by God's word. Now there is no difference except that Adam came from the earth, but Mary was not created from the earth' '*aus dem mund gottes ist Adam beschaffen. nun ist da kein undterscheid als allein, daß Adam aus der erden ist, Maria aber nit von der erden beschaffen gewesen*'.[64] Mary was not created from earth, which means, following his reasoning from the first part of *De genealogia Christi*, that she had to be eternal. Furthermore, Mary was created first, 'Mary was the first among all other creatures, the first born, and her son was the first among men. So Mary came from God's word, not from the earth, rather she emerged from the heart of divinity' '*Maria ist die aller erst undter allen creaturen gewesen, die erst geboren, und ir son der erst geboren undter den mannen. also ist Maria aus dem mund gottes gangen und nit aus der erden; nit allein aus dem mund gottes aus-*

gangen, sondern aus dem herzen der gottheit'.[65] According to Katharina Biegger, Paracelsus situates Mary in the Trinity because of the way she was created.[66] He likely formulated this construct because he could not fathom a holy figure being born from a womb that was entirely human. Paracelsus goes so far to suggest that Mary was never completely human in form: 'That is: through Christ, her son, she covered the earth. For the knowledge of God became her flesh and blood and a human that is the son of God"*das ist: durch Christum, iren son, hat sie die ganz erden bedeckt. Dann die weisheit gottes ist ir fleisch und blut worden und ein mensch, das ist der son gottes*'.[67] He continues, '[Mary is] not like other women, who go forth in the bright abyss, who are born after a long time, who know nothing about beginning. She is the first and has no end' '*nit als andere frauen, die in abgrund der hellen fahren oder die, so geboren werden nach langer zeit hernach, die nichts von anfang wissen. Sie ist die erst, und hat kein end*'.[68] Mary does not have a human body, rather she resembles an in-between kind of creature not unlike the elemental spirit (a concept Paracelsus uses in other texts).[69] Not human herself, she mediates the action of God.[70]

Conclusion

In the tracts considered in this chapter, Paracelsus chooses to understand Mary as an in-between being (between humans and divine beings) and as an almost invisible, transient shell of a woman. His thoughts about Mary confuse his stance as an expert on human birth and reveal his fear that perhaps human birth is not directly willed by God. Because he thinks that all humans must emerge from a divine seed, he expects that Christ's birth must be more special because of Mary's body, not necessarily because of her her virginal status. He entirely overlooks the idea that Christ's birth was a virginal birth. Instead, the underlying assumption in these tracts was that Mary's body must have been physiologically different than other female bodies. His core assumption that something must have been special about Mary's body did not yield the expected result of explaining the process of conception in a way that adequately underscored the significance of Christ's birth. Paracelsus repeatedly appears to be uncomfortable with the notion that there may be a connection between the Mary and womb, though he is unable to define it.

His consideration of Mary in the birth of Christ and his wish to distance Mary's body from that of all other females demonstrates his concern that women may have special power or influence over the material in the womb. Though he still does not come to a conclusion about the efficacy of the human womb in his mariological tracts, he does achieve some success in differentiating between spiritual and corporeal births when he writes of the metaphorical second birth that results from the belief in Christ. However, since he considers the metaphorical birth to be predestined and ultimately tied to the corporeal birth of humans, he is left to consider the degree to which God influences the creation of a human body. He works through this conceptual challenge by thinking about the meaning of Adam and Eve in what are considered his medical tracts, as the next chapter will make clear.

4 ADAM, EVE AND THE HUMAN BODY: PARACELSUS'S NATURE DILEMMA

Paracelsus uses the birth of Adam and Eve to help him define the natural foundation of the human body and the divine origin of the soul in the following tracts. As he thinks about the significance of Adam and Eve, Paracelsus expresses trepidation when he begins to envision the seed of human conception as possibly functioning separate from the divine. His unease in categorizing the seeds of conception as entirely part of the natural world stems from his vociferous lament against what he called 'foolish doctors' in many of his other tracts on birth. Recalling that he argued against the kind of medical practice that sought to manipulate human conception, his thoughts about the natural world could undermine his claim that he offers a better approach to medicine. When he argues for the soul as a body within the human body this helps him regain his footing. It also facilitates for him a way to understand that the birth of Christ was more significant than other human births. This was a problem he wrestled with in his Mariological tracts where he aimed to understand human birth as different from Christ's. In the Mariological tracts he focused on Mary's body to understand why Christ's birth was more significant than other human births and was unable to clearly delineate the features of human physicality that differed from Christ's. However, thinking about the bodies of Adam and Eve offers him an opportunity to envision divine and natural aspects of human bodies. When he contemplates the limits of God's work in their births, he begins to articulate an idea of the physical body as separate from the soul and spirit. Furthermore, the comparison of all humans to Adam and Eve leads him to recognize physiological differences between men and women. Adam and Eve serve as models for understanding human bodies in a way that contemplating Christ's birth could not. This chapter explores the various (and inconsistent) ways that Paracelsus contemplates the origins of the body and soul in order to demonstrate how reflecting on the figures Adam and Eve helps him clarify his theory of embodiment. Because his writing takes many twists and turns as he works through this issue, the chapter will explore one tract at a time in order to clarify how his understanding of humans, Adam, Eve and nature are articulated in the specific tracts.

When Paracelsus initially wrote about conception he maintained his stance against the usefulness of Aristotle's works in the field of medicine. He continues to condemn classical authority in the works considered below. Yet, when he enters into reflection on the meaning of the soul he positions himself within a philosophical tradition that could, if his competition noticed it, realign him with Aristotle. As the earlier chapters demonstrated, Paracelsus vehemently objected to physicians who relied on Aristotle's understanding of the human body. Yet, Aristotle articulated the idea of the soul as a body: an idea that Paracelsus likewise champions. Among early philosophers the soul was understood as the animating principle of the body. Later, Plato combined the religious and philosophical notions of the soul, so that the soul also acted as a moral agent. For Plato, the soul was incorporeal and thus also immortal. Aristotle was less specific about the soul and his writing left room for debate about whether or not the soul could be immortal. In the thirteenth century and later, however, Aristotle became the chief philosophical authority. The *Council of Vienne* declared the Aristotlian definition of the soul as a form of the body to be an article of faith. However, discussions of the immortality of the soul seem to have waned until they resurfaced in the fifteenth and sixteenth centuries.[1]

In the first half of the sixteenth century, the nature of the human soul was a major controversy in Italian academic philosophy, precisely the time when Paracelsus was studying in Italy. The two principal participants in the discussion in Italy were Pomponazzi, who maintained that the authentic Aristotelian view that the human soul was mortal, and Agostino Nifo, who argued the notion of an immortal soul could be upheld on the basis of ancient authority.[2] It is plausible to consider that Paracelsus encountered discussion about the nature of the soul during his time in Italy. But he used his discussions of the origin of the soul to underscore his special place among medical practitioners in German-speaking regions. Thinking about the origins of the human soul took on both religious and medical significance in Paracelsus's tracts. The works of later German physicians suggests that he gained a following because research demonstrates that using metaphors of Adam and Eve to discuss human birth and the meaning of the soul unfolds in later German medical and theological writing.[3]

Paracelsus uses his interpretation of Adam as a framework to define the origins of the human soul and delineate the physical qualities of the human body. This inquiry emerged from his longstanding quest to understand the process of conception and to define his place in the field of medicine as evident in his references to his own theory of conception. This is not the way others have dealt with Paracelsus's engagement with Genesis. More often than not, Paracelsus's writing on Adam tends to be interpreted as part of his theological interests and research on the following texts tends to compare Paracelsus to Lutheran readings of Adam.[4] While Paracelsus was likely influenced by his contemporary,

Martin Luther, his writing on Adam and Eve also requires understanding of how it fits into his larger quest to be known as a physician and to redefine the field of medicine. I argue that it is the culmination of a longstanding quest to define the human body in a new way that leads Paracelsus to develop his theory of embodiment by comparing and contrasting the bodies of humans, Adam and Eve. Because of his compulsion to define humans as spiritual beings, he finds it necessary to investigate how human physicality relates to the divine.

Bruchstücke des Buches Von den Fünf Entien, Gennant Volumen Medicinae Paramirum de Medica Industria [*Excerpts from the Book about the Five Entien, called Volumen Medicinae Paramirum*] (date unkown)[5]

Volumen Medicinae Paramirum, represents one of Paracelsus's early contemplations of the human body as it relates to Adam and Eve. In this tract he concentrates on the spirit and suggests that the physical body of humans differs from the spiritual body.[6] *Volumen Paramirum* builds upon Paracelsus's quest to define himself as a different kind of doctor and his exposition on the spiritual qualities of humans underscores this new approach to the practice of medicine. Like other early texts, Paracelsus seeks to establish his authority by undermining the approaches of his competition, yet in this tract he views erudite physicians as his greatest rivalry. He explains that there are several different approaches to the practice of medicine based on the works of Aristotle, Hippocrates, Galen and Albertus Magnus and he discredits all of these approaches.[7] He wishes to reassert his religious-based medicine as superior to other practices that do not take the divine into account. In *Volumen Paramirum* he takes a stab at the writing of those practitioners who are not Christian, when he explains his preference for writing in 'Christian style' (*christlicher stylus*), a style exemplified by belief in Christ.[8] Although there are similarities to other early works the way that Paracelsus positions himself in the field of medicine, his understanding of the human body and its initial conception differs significantly.

Unlike other early works, he rejects the postulate that humans are divinely formed and instead explains reproductive matter as a natural entity. He writes, 'So you say that heaven makes bodies, which is the *astrum*, but it is not. Humans are physically created once and after that they are formed from the *ens seminis* alone without all the stars' '*also ir sagt, das der himel den cörper macht, das ist das astrum, welches nit ist. der mensch ist einmal beschaffen corporaliter und weiter formirt in nichts dan alein ens semins on alle gestirn*'.[9] Unlike the other earlier tracts discussed in Chapters 2 and 3, Paracelsus precisely defines the limits of divine involvement in creating the human body. In *Volumen Paramirum*, he emphasizes the independence of the *ens seminis* from the divine when he explains that the *ens seminis* and God created Adam and Eve, but that all other humans were made from the *ens seminis* alone.[10] In other words, the formation of Adam and Eve was a combination of divine power and natural entities, but that was the last time God intervened in human birth. This categorization of humans as created differently than Adam and Eve with a sharp focus on natural qualities in humans – the human seed – signals a shift in Paracelsus's thinking. He conjectures that the natural world might be different from the auspices of divine power and in this way he defines an aspect of embodiment that distinguishes the human body from the divine, namely the physical qualities of humans may lack a divine connection. He distances human birth from the birth of Adam and Eve further in a section that Huser (and later Sudhoff) identified as being in his own handwriting:

> If we want to talk about the initial formation of the body, we observe that Adam was materially created and thereafter nothing else was created. That which is conceived in the mother's womb, flesh and blood, a body that becomes a complete human – that occurs through a natural *ens* mediated by parental seed. A seed thrown in a field, now that grows from its own seed and with no other.
>
> *dan so wir wollen vom anfang formirung des cörpers reden, so betrachten wir, das Adam materialisch geschaffen ist und weiter nichts mer geschaffen wird. was aber im mutterleib empfangen wird, fleisch und blut, ein cörper zu einem volkomen menschen, das selbig geschicht aus ens naturae mediante semine parentum. ein samen der in ein acker geworfen, derselbig der wachst alein aus dem eins sein selbs und keins andern.*[11]

The qualification that the body is formed through a natural *ens* mediated by parental seed ('*aus ens naturae mediante semine parentum*') reflects a change in Paracelsus's thinking. In this passage it is clear that he considers the process of conception to be influenced by an outside force, the *ens naturae*, even though the parents' seeds (*samen*) ultimately determine how the body of the child will form. Unlike most of his other texts on the process of birth, he does not suggest that God is directly involved in the birth of humans, in the formation of the body; rather, he implies that the forces of nature determine the physical characteristics of the human.[12] Even though in several of the tracts explored in previous chapters he viewed the outside force as divine, in this tract he suggests that the external influence on the creation of a human body may not have heavenly origins.

In *Volumen Paramirum* Paracelsus expresses his conviction that the physical body, but not the soul, should be considered part of the natural world. He articulates this point by explaining the physical body of a mother and its relationship to an unborn child: 'a child will not grow without the abdomen, for it grows in the abdomen, that is, in the womb, and therefore no stars or planets influence it: his mother is his planet and star' '*ein kint mag on die digest nicht wachsen, dan im digest wachst es, das ist in matrice; und also bedarf das kint keins gestrins noch planeten darzu: sein muter ist sein planet und stern*'.[13] While he draws parallels between the mother's body and the natural world, he still seeks to explain both the human physical body and the connection between the natural world and the divine. He understands the womb to be a microcosm of the natural world, a miniature of God's created universe that nonetheless remains separate from divine influence.[14] Yet, he also maintains that, although the mother carries the child in her womb and acts as a 'star and planet' for the developing foetus, her body is merely a container that nourishes the child. It is a physical property that lacks divine connection.

He underscores this correlation between the womb and the natural world further when he explains that the child lacks a spirit (*Geist*) until the moment of birth. Paracelsus reasons that this separation of the body and spirit prevents the being growing within the mother from being human. He assumes that to be human means to have both physical and spiritual qualities. He reasons that the spirit provides a connection to heaven, a bond that he regards as unique to humans. However, he finds it difficult to separate the body and spirit entirely. Although he emphasizes that the spirit is different from the body, he asserts that it is connected to the body: 'so the body and spirit are united, for the spirit comes from the body through one's will' '*nun die cörper und die geist sind vereiniget, dan die geist komen vom cörper durch die willen*'.[15] In an attempt to explicate how the spirit could develop without a body, he implies that the spirit grows once the child wills it to be there. In other words, it is the child's conscious decision to receive the spirit.[16] Whether or not the ability to develop this faith is part of conception, as he maintained in *De Geneologia Christi* (see Chapter 3) he does not explain. He does, however, clearly demarcate the idea that humans are dualistic and should be understood as two bodies in one, or a spiritual body and a physical body. His attempt to define the physical body as different from the spiritual body leads him to define the natural realm as distinct from the godly domain.

The way that Paracelsus works through his understanding of the dual qualities of humans: spiritual and physical, reveals the interplay between his longstanding contemplation of human birth and his quest redefine his role as a physician. On the one hand, he signals that physicians must be aware of the dual qualities of humans and on the other it becomes clear how his fascination with the process of conception serves as a model for understanding the workings of the body and the origin of spiritual qualities. For example, he begins to envision the mind working in a corporeal way when he uses a metaphor of giving birth to describe the origins of the spirit, 'This culminating desire is a mother that bears the spirit. Understand it this way: Thoughts make a word and the thoughts are a mother of speech, so where there is no thought there is neither speech nor words' '*Diser beschloßner wille und verhengter ist eine muter die da gebiert den geist. Also zu verstehen: die meinung macht ein wort und die meinung ist ein muter des redens, also wo die meinung nit ist, da ist kein red noch wort*'.[17] Using the metaphor of human birth to explain the workings of the body recurs in his later works and is a trend that is most pronounced in the tracts in which Paracelsus also considers the meanings of Adam and Eve. He begins to divide his comparisons. Thinking about the birth of Christ relates to spiritual aspects of human existence, whereas the reflections on Adam and Eve are tied to both a contemplation of the natural world and the human body within that world.

In some segments of *Volumen Paramirum* Paracelsus conflates spirituality with physicality. He defines the emergence of faith as a kind of spiritual birth and begins to envision spirituality in a physical way. However, the tract also reveals tension between understanding the body as a natural phenomenon that has limited connection to the divine and understanding spirituality. While Paracelsus has difficulty in clarifying his understanding of spirituality as different from physicality in *Volumen Paramirum*, using Adam and Eve as reference points he works through this conundrum in later works. In *Volumen Paramirum* Paracelsus approaches the study of humans' spiritual qualities from the framework of a physician attempting to understand the physical workings of the body. He also uses Adam and Eve as guideposts to understanding the dual nature of humans in what many consider his magnum opus, *Astronomia Magna.*[18]

Astronomia Magna oder die Ganze Philosophia Sagax der Großen und kleinen Welt (1537/8)[19]

Astronomia Magna clarifies his thoughts on difference between physical and spiritual qualities by contemplating limits of divine influence. It also reveals Paracelsus's suspicion that nature itself may be a force upon the physical body, but this construct is only comprehensible in the context of Paracelsus's other writing on human birth as it relates to Adam and Eve. Whereas in earlier tracts Paracelsus sought to explain human birth by comparing it to Christ's, in *Astronomia Magna* he focuses on the difference between Christ and humans. When he considers the disparity between the body of Christ and the human body he considers that humans must be compared to Adam and Eve. Paracelsus explains, 'The first, as the father created humans from below. The other, as the son, from above' '*der erste, als der vater, hat geschaffen den menschen von unden herauf. Der ander, als der son, von oben herab*'.[20] This sentiment marks a shift in his thinking. Unlike earlier works where he sought connections between the birth of Christ and the birth of all humans, here he aligns humans with Adam, reasoning that humans have some special connected to God while Christ originates entirely in God's realm.

Like Adam and Eve, humans form in the *firmament*. Using the analogy of Adam and Eve helps Paracelsus delineate natural and spiritual qualities. He does not abandon the idea that humans must maintain some divine connection, though he does doubt that the connection is entirely physical. Because he seeks a divine connection for humans, Paracelsus suggests that humankind is both mortal and eternal, 'since the human his twofold, in one part mortal and the other part eternal' '*so doch der mensch zweifach ist, in einem teil tötlich im andern ewig*'.[21] Paracelsus thus understands humans to have two bodies: one is spiritual and one is physical or elemental.

His longstanding contemplation of human birth influences the way he articulates the presence of the two bodies within humans: 'It is like when a man and a woman together give birth to a child and without both nothing will happen, therefore two lights were given to us to dwell in one human' '*nun ists doch gleich als ein man und weib die beide ein kind geberen, und on beide mag es nicht beschehen: also seind uns auch beide liechter geben in einem menschen zu wonen*'.[22] The metaphor of conception provides him with a model of the union of corporeality and spirituality. In order to explain how the spiritual half of humans develops, he considers the mind, but first he redefines the relationship between the human body and God.

In the *Astronomia Magna,* Paracelsus focuses on the likeness of humans to God, distancing the body and reason from the divine and, instead, regarding only human likeness as being closest to God. This logic calls to mind *Buch von der Gebärung* where humans were seen as most like God because Christ chose to be a 'two-footed' creature (like humans). However, in that tract Paracelsus took God's choice of placing Christ in a human body to mean that humans were a superior creation conceived with divine assistance. In the *Astronomia Magna,* only cognition emanates directly from God and, likewise, leads back to God.[23] For Paracelsus, however, separating the divine from the elemental, the body, presents a special challenge. He suggests that God gave humans the ability to reason so that they could know Christ and learn from the stars. Emphasizing the relationship between cognition and the astrological sphere he writes,

> Therefore human reason has a magnet that draws from the stars reason and thoughts. Therefore the human is placed in two bodies, the one is visible and the other invisible, that is in the elementary and the heavenly. So notice further that the body comes from the elements, the spirit from the stars.
>
> *also die vernunft des menschen hat einen magneten, der in sich zeucht vom gestirn die sinn und gedanken. Also ist der mensch gesezt in zwen leib, das ist in den sichtbaren und unsichtbaren, das ist in den elementischen und himlischen. Darauf merkent iezo weiter: der leib komet aus den elementen, der geist aus dem gestirn.*[24]

The visible and the invisible bodies represent corporeality and spirituality respectively. However, it remains unclear whether Paracelsus groups reason (*vernunft*) with spirituality in the invisible body. Working with the notion of the body as space and the undetectable realm of the spirit, he continues, 'for blood and flesh live in the body and reason and thoughts reside in heaven' '*dan im leib wonet blut und fleisch, im himel wonet sinn und gedanken*'.[25] While he postulates that God participates in the formation of the body, he encounters difficulties when relating his thinking about what we understand today to be the mind, to his ideas about the body. He regards reason, itself an invisible body, as another part of humans that is enclosed in the body.

Having distinguished the mind from the physical body, he considers whether and how the visible body might be associated with the divine. He returns to the birth of Adam and Eve, this time postulating a difference between Adam, Eve and all other humans. He convinces himself that humans were made in the image of God and encompass in their being the entire universe. He maintains that God only wanted to create one human: 'and because God did not want to pull two people from the world, rather only one, one that is the fifth being, the microcosm, he made the man, not the woman'. '*und alein das got nicht hat wöllen zwo personen aus der welt ziehen, sonder nur eine, der das fünft wesen ist, und der microcosmus, den mann ausgezogen, nicht die frau*'.[26] Paracelsus thus asserts that God created only the male body (Adam), but this leads him back to his thoughts on human birth. Because he had so long maintained that human conception was divinely willed, he comes up against his own theory of human birth when he tries to rethink the roles of men and women based on the idea that God created only Adam.

Though he asserts that two humans are necessary for reproduction to occur, that Eve was made from Adam and men and women belong together and cannot live without each other, he envisions different roles for men and women in reproduction. He now states that God is not directly involved: 'That is why when I write about children they are not further made by the hand of God, rather they are ordered to be born in by spirit of nature'. '*darumb von den kindlin zu schreiben, so weiter nicht gemacht werden von der hant gottes, sonder in die geschöpf der natur befolen, geboren zu werden*'.[27] He explains that God's direct participation in creating a human was limited to the birth of Adam. After Eve was born from Adam, nature organized reproduction and brought order to the world.[28] Attributing any further human conception to nature marks a significant shift in Paracelsus's thinking that retracts theories put forth in his other writing. He models human conception on his understanding of Eve growing from Adam. That is, that the male possesses the material for reproduction and that all humans grow from the substance that began in the male body. He therefore confuses his previous assertions about the process of conception, while at the same time clarifying his thoughts on the limits of the physical body.[29]

The association of humans to Adam facilitates for Paracelsus a way to consider the birth of Christ as different from humans, a problem that he sought to explain in many earlier works. The idea that humans formed from below, like Adam, leads him to articulate a sense that things occur differently in the natural world, than beings created directly by God. Paracelsus goes so far as to describe the birth of Christ as having occurred beyond nature, or supernatural: 'Mary carried that which the heaven and earth made, that which is more than all the world and not like nature, but more than nature' '*Maria hat getragen den, der himel und erden gemacht hat, das vil mer ist dan alle welt und nicht nach der natur, sonder uber die natur*'.[30]

Using nature as the defining characteristic of human birth enables Paracelsus to distinguish between divinity and corporeality. In addition, separating the birth of Christ from the influence of nature enables him to clarify his view of nature. In other words, he comes to recognize nature as distinct from the forces emanating directly from God. This becomes an important component in Paracelsus's understanding of wondrous or monstrous births. He explains that the birth of an incomplete human ('*kein volkomener menschen, sonder ... ein misgewechs*') must be blamed on nature, which includes the stars. Despite their incompleteness, Paracelsus regards such beings as human. This was not the common understanding of birth defects or unusual physical traits at the time.[31] He seeks what could be called a scientific basis of birth defects, rather than relying on the common belief that a Satanic, astrological or divinely-willed force caused deformity. In this way, his theory of the body as part of a natural world suggest that he feels compelled to distance physicality from supernatural forces.

His uncommon theory of the body in this tract undermines his own long-held assertion that the body was within God's realm. In addition, the *Astronomia Magna* reveals Paracelsus's confusion about the forces acting upon the physical body. He does realize that humans have both spiritual and physical qualities and seems to more easily define the spiritual qualities because of the direct connection to the divine, but in the end his thoughts on the physical body leave him baffled. He even expresses a threefold embodiment in *Astronomia Magna*, and this is a constellation that he seldom addresses in other tracts. In *Astronomia Magna* he refers to the soul (*Seele*) and the spirit (*Geist*), whereas in several other tracts the soul and spirit were used interchangeably, rather than as separate aspects of the human. He does, however, continue to recognize the bipartite qualities of humans: body and soul; yet at times he hints at a tripartite version of human embodiment: body, soul and spirit. These expressions all reveal a quest to understand the physical body as distinct from the spiritual aspects of humans.

Philosophia Sagacis

Paracelsus begins *Philosophia Sagacis*, the second half of the *Astronomia Magna*, by stating that what the reader finds in the tract will be more theology than natural philosophy.[32] However, though he claims to focus primarily on theological issues, the tract puts forth a more comprehensive overview of his ideas about nature than the one he offered in the *Astronomia Magna*. His exploration of the birth of Christ and Adam in *Philosophia Sagacis* leads to a more concrete concept of human corporeality.

Building on the notion of the dual nature of human bodies presented in the *Astronomia Magna*, Paracelsus claims: 'therefore there are two parts in humans: from the earth the form, from God the spirit' '*also sind zwei stück da im menschen: von der erden die form, von got der geist*...'.[33] His definition of humans as physically related to the earth and spiritually related to God is reminiscent of the *Astronomia Magna*. But Paracelsus continues to work through the idea of nature existing separate from God and focuses on the origin of the spirit, 'therefore God made humans from the earth first, thereafter the spirit was blown into them through the spirit of life, through which the human is capable of ascending, without going back to where he originated' '*also hat got den menschen am ersten von der erden gemacht, darnach den geist einblasen durch das spiraculum vitae, durch welchen der mensch fehig ist wider hinauf, on dannen er komen ist*'.[34] Recognizing that the spirit (*Geist*) is the bridge that connects humans and God, Paracelsus delineates the human qualities that he deems natural and those that are divine. Whereas he had previously focused on birth to aid in defining the human body, he now turns to death: 'therefore what is from nature will return to nature and to God what is from God. For we should recognize that everything that is from the earth will not ascend, only that which is from God will ascend' '*also kompt in die natur was von der natur ist und zu got was von got ist. wan wir sollen das wol merken, das alles was von der erden nit auferstehen wird alein was von got ist. wan wir sollen das wol merken, das alles von der erden ist nit auferstehen wird, alein was von got ist, das wird auferstehên*'.[35] Paracelsus reaches his greatest clarity on the qualities of human embodiment in this tract: he carefully delineates the earthly qualities in the physical body and the spiritual qualities in the soul. Furthermore, in this tract he differentiates between the spirit (*Geist*) and soul (*sêl*). In other tracts he had used the terms interchangeably and never spelled out exactly what he meant by them. Here he explains that it is important to understand the birth of the soul as different from the birth of the spirit. The spirit was given to humans by God, 'it is not called a soul, that is the spirit, that is blown into humans after they are formed, namely life and the spirit of life' '*so heißt es nicht ein sel, das ist, der geist, der nach der formirung des menschen eingeblasen ist von got, nemlich das leben,und ist spiraculum vitae*'.[36] This constellation of the moment a human gains a soul squarely places conception and the physical body in the natural world and suggests divine intervention only in the latter stage of pregnancy. He describes the birth of the soul thus:

> Now the birth of the soul is as follows. The child is conceived in the flesh, in other words born, in seed, a word is spoken by God in this conception in the flesh, God's word gives the body a soul. Now be aware that the soul is the centre of humans, in which all other spirits now live: good and evil.
>
> *Nun ist die geburt der sêl also. So das kind entpfangen wird im fleisch, das ist geboren, im samen, in der selbigen fleischlichen entpfengnus gehet von got aus ein wort, das gibt dem fleisch ein sêl. nun merkent, die sel ist das centrum des menschen, in welchem alle andere geist iezt wonen, gut und bös.*[37]

This description of the origin of the spirit parallels his thoughts in *Volumen Paramirum*, in which the spirit became part of the human after birth. The spirit is breathed into humans after the body is formed and by contrast God gives the soul to humans at the moment of conception. Paracelsus seeks to distinguish the qualities of human birth that are divine and those that are natural.

Although in *Philosophia Sagacis* Paracelsus asserts that God is present at the moment of conception, Paracelsus implies that the seed (*samen)* is different and separate from the part of humans that develops with divine help. In this constellation, God is involved only in the spiritual conception, when the soul is placed in the body. This differs greatly from placing the reproductive materials into the human, as Paracelsus had described in *Buch von der Gebärung*. In the *Philosophia Sagacis* the soul, placed in the human by God at conception, prepares the body to receive good and bad spirits when the physical body is complete. Thus, Paracelsus devises a way to separate body and soul, considering them two entities within the same human body. The soul allows both good and evil influences to affect the body. This suggests that harmful/bad (*bös*) spirits enter the body through a pathway (*sêl*) created by God. In order to distance evil from God he explains that nature is involved in making the soul physical, 'You should know what the soul is. It is the breath from God's mouth, and through nature it is turned into flesh, mortal flesh, which is created through belief, [it] becomes living flesh of new birth' *'also wissent, was die sel ist, das sie ist ein atem vom munde gottes, und ist durch die natur gefleischet in das tötlich fleisch, welchs geschaffen ist in den glauben, zum lebendingen fleisch der neuen geburt'*.[38] This confusing passage demonstrates his confliction between thinking about some aspects of human physicality as tied to nature alone, and some directly created by God. It reads as if he is trying to situate Christians (what he refers to as the new birth) in relation to the divine and to nature, but his goals are entirely unclear. However he does seem to be attempting to distance bad human qualities from the divine and it is in this vein that he articulates a sense of the natural physical body as separate from the divine. When he attributes harmful qualities to the influence of nature he saves himself from implicating God. This leads Paracelsus to assert that part of the body is connected to nature, and the other is connected to God. However, because of his consistent belief that humans have a special relationship to Christ and to Adam, he turns to divine births to define human corporeality.[39]

Paracelsus classifies the two bodies as mortal (tötlich) and immortal (untötlich). The body as it relates to Adam is mortal. Adam became mortal because he was made of the earth and must return to the earth, which solidifies the idea of human corporeality as related to the earthly realm. In *Philosophia Sagacis* Paracelsus adds that Adam would have returned to God had he not sinned.[40] Thus, Paracelsus moves to distance immortal bodies from Adam focusing specifically on Mary and Christ:

> For Adam's body should be nothing. And in the beginning it was as follows: The new birth comes from the virgin and not from women. It follows then that the virgin from whom the new birth originates, was a daughter of Abraham as it was commanded and not from Adam, that is, she was born from Abraham without human seed, the power of God's word without any kind of mortal nature. Christ was born from the virgin, who therefore was not from Adam, not from his seed, nor from his flesh. He was conceived from the Holy Spirit and became incarnate not in the order of mortal bodies, rather like a new birth that emerged from the Holy Spirit.
>
> *Dan Adams fleisch sol nichts. Und anfenglich also: aus der jungfrauen gehet die neue geburt und nicht aus der frauen. Aus dem folgt nun, das die jungfrau, aus der ausgangen und geboren ist die neu geburt, ein tochter ist gewesen von Abraham nach der verheischung und nicht aus Adam, das ist sie ist gewesen von Abraham on menlichen samen geboren, in kraft der verheischung on alle tötliche natur. Aus der jungfrauen, die also von Adam nit ist, nit von seinem samen; nur aus seinem fleisch, ist geboren Christus, der empfangen ist vom heiligen geist und vom heiligen fleisch incarnirt nicht nach der ordnung des tödliche fleischs, sonder nach der neuen geburt, die da gehet aus dem heiligen geist.*[41]

His understanding of Mary's divinity reflects some of his thinking on Mary's birth put forth in his Mariological tracts.[42] However, in *Philosophia Sagicis* Paracelsus uses the birth of Christ to expound upon his notion of human corporeality and spirituality. He equates the uniqueness of the birth of Christ to human's spiritual qualities. Only Christian bodies, those granted a spirit by God at the time of conception, can receive Christ.[43] Paracelsus explains this phenomenon in terms of the body of Christ and Adam: 'Therefore Christ was not from Adams body, only from the flesh and blood of heaven, for through Christ's birth, he is the original, the first born in the new birth' *'also mocht auch Christus nichts ausrichten durch Adams fleisch, alein durch das fleisch und blut, das von oben herab war, das dan durch Christum geboren ist, in welcher zal Christus primogenitus ist, der erst geboren in der neuen geburt'.*[44]

When he distinguishes Adam's body from Christ's Paracelsus signals that he sees the book of nature as being independent of God.[45] Furthermore he underscores the dual nature of humans, 'therefore there remain two kinds of knowledge in humans, namely the natural and the heavenly' '*also bleiben zwo weisheit im menschen, nemlich die natürlich und die himlich*'.[46] Paracelsus comes to an understanding of the varying roles of nature and the divine by interpreting the relationship of God to Christ and God to Adam through their respective births. As he continues to read Adam's body as representative of the natural body, he considers the possibility of demonic influence on earthly bodies, 'for the newborn flesh cannot be ruled by the devil, only Adam's body can be taken by the devil wherever he wants' '*dan das neugeboren fleisch regirt der teufel nicht, alein Adams fleisch mag er bringen wohin er wil*'.[47] He underscores the vulnerability of the natural body, a body that exists without a spiritual half.[48] In other words, only the embodied person, one with natural and spiritual qualities can be protected by God. Thinking about Adam helps Paracelsus arrive at this conclusion. When he considers Eve he recognizes the natural body sexed, but such thinking does not arise when he contemplates only Adam.

Liber de Matrice, a portion of *Opus Paramirum* (1531)[49]

Paracelsus's contemplation of the physical body as it relates to Adam and Eve in *Liber de Matrice* underscores the way his philosophy of the body reveals a crossroads between medieval and modern thinking about the body. Initially, he writes about the close connection between Adam and Eve, noting that Eve began her existence as a man because she was part of Adam's body:

> Just as the fish is different than water, and the water different from the fish, it is therefore necessary that doctors know this likeness, that he is a seed and the woman is a field and although she is also a seed, she is from the man. She does not remain a man, but rather becomes a woman. For when she was a man in Adam, once the hand of the creator came she was no longer a man, no longer of Adam's rib, therefore you need to understand and recognize that a different kind of medicine applies to her, one with a different theory, one requiring observation.
>
> *So nun dem fisch ein anders zustehet als dem wasser, dem wasser ein anders als dem fisch, so ist nun hierauf bilich, das der arzt solch gleichnus merke, das er ein sam ist und die frau der acker und wiewol sie auch ein sam ist, aber aus dem man. Sie bleibt aber nicht ein man sonder sie wird ein frau, ietzt ist sie dem samen entrunnen und nimer nach dem samen zu urteilen. Dan wie sie ein man war in Adam, aber da sie in des werkmeisters hand komen, da war kein man mer da, kein Adams ripp mer; darumb so ist bei inen das zuverston und zuerkennen, das sie sollen vom arzt in ein sondere monarchei gefürt werden mit einer andern theorica und physica und die augen baß auftun.*[50]

The idea that Eve emerges from Adam as a version of him reveals what Laqueur called a one-sex body and also parallels medieval thinking about the physicality of

Christ as androgynous.[51] But, even though Paracelsus focuses on the emergence of the female form from a male form, he explains this as a one-time occurrence that was facilitated by God. He emphasizes that since that time male and female bodies differ. Paracelsus takes the relatedness, yet difference, of men and women to explain the need to understand that women's illnesses require a special kind of medicine.[52] He suggests that there are physiological differences between the bodies of men and women that require physicians to treat the female body differently to the male body. He further assumes that their physiological differences must correlate to their different roles in reproduction. In other words, he views the birth of Adam and Eve as a kind of metamorphosis akin to what Laqueur described as a one-sex model of the body, but emphasizes the need to recognize that humans, unlike Adam and Eve, need to be understood as two kinds of humans and need to be cared for based on their sex. Thus, he views male and female physicality as inherently different, or to use Laqueur's terminology again, he posits a two-sex model of the body.

Although distancing the birth of Adam and Eve from all other humans clarifies his medical perspective on the treatment of men and women as resolutely different, his contemplation of Adam and Eve as a metaphor for human birth confuses his previous assessment of the process of conception and leaves him repeating a kind of Aristotelian model that he previously eschewed. He conjectures that since the time Eve separated from Adam, women have no influence on seed (*samen*). Through Adam, men retain their closeness to God, but women, as emphasized by their limited role in reproduction, are one step removed from divinity. That is, the male determines when and how conception will occur. This theory of conception sharply differs from previous tracts. His interpretation of the Fall informs his understanding of the process of conception and forces him to rethink previous assumptions about the female body and the workings of the womb.

In *Opus Paramirum* Paracelsus ascribes a different role to the womb from what he had in the previously discussed tracts. Here, the womb receives and nourishes the reproductive material, suggesting that the male seed (*samen*) contains all necessary elements to bring forth new life. When Paracelsus removes women from the initial stage of reproduction, he redefines the woman's role in birth following an Aristotelian model.

Indeed, he moves away from understanding conception and gestation using biblical metaphors and instead uses metaphors of the female body to describe the natural world and biblical figures therein. He describes the creation of the world using a metaphor of the womb, linking the human womb to Creation:

> Therefore there are three matrices: the first the water that carried the Holy Spirit, and that very matrix was made into heaven and earth. Thereafter heaven and earth became

> the matrix for Adam, who was created by God. From the man, woman became a womb for all humans until the end of time.
>
> *Also seind drei matrices: die erste das wasser, auf dem der geist des herrn tragen ward, und daselbe war die matrix in der himel und erden beschaffen worden. Darnach ward himel und erden in die matrix Adae, der durch die hant gotes gemacht wurd. Und aus dem man ward die frau ein matrix aller menschen bis in das ende der welt.*[53]

Paracelsus's circuitous argument, namely, that on the one hand, women lack a direct connection to God that men retain through Adam, but on the other hand that there exists a relationship between Creation and the womb, leads him to connect the womb with the natural world.[54] Throughout *Opus Paramirum,* Paracelsus emphasizes that like the original matrix that created everything, the womb from which humans are born remains invisible.[55] The three invisible wombs differ, as the above passage makes clear. The spirit, in this case he means the Holy Spirit, was present in the first matrix. When the second matrix was formed to make Adam, God was directly involved in his birth, which means that God was only directly involved in the creation of Adam. Through Adam, who was part of the original matrix, the human womb carries the Holy Spirit along with the child.[56] In this constellation of human reproduction, the Holy Spirit is present in conception, acting as an intermediary between God and humans, rather than being directly involved in reproduction, as Paracelsus had previously understood. Paracelsus posits a divine connection in the womb; but rather than directly and actively participating in reproduction, Paracelsus regards the Holy Spirit as a presence.[57] Even though he suggests that a divine power resides within the womb, he still considers the possibility that the female body influences the womb in a way that is independent of divine influence. His reflections on the working of the womb suggest that he views physical aspects of humans as tied to the natural world, and in particular the female body is closer to the natural world, the male body closer to the divine.

Paracelsus articulates his conviction that the physical body should be compared to the natural world in his formulation of the difference between humans, Adam and Eve and also by associating the womb with the natural world. For example, when Paracelsus compares the womb to a tree that takes nourishment from the earth, he surmises that the womb is dependent on the body for its health: 'if her body is not good, unhealthy, not in the right order, the womb is spoiled, infertile, unhealthy and filled with all kinds of other illness' '*so ir leib nicht gut ist, nicht gesunt, nicht in rechter concordanz, so ist alle matrix verderbet unfruchtbar, ungesunt, entschickt und mit allen andern zufallenden krankheiten beladen*'.[58] In this tract, Paracelsus fluctuates between understanding the womb as comparable to nature and nature as comparable to the womb. While he lacks clarity in defining human physicality and the natural world, this way of explaining the relationship between a woman's body and her womb leads Paracelsus to delineate the need for a special approach to women's medicine.[59]

In the *Opus Paramirum*, he distinguishes between male and female illnesses locating the distinction in the uniqueness of their bodies.[60] Paracelsus contends that physicians should judge from the physical appearance how to treat the patient.[61] Regarding the female body as more closely related to nature, he writes, 'woman is closer to the world than man and man is farther from her in the anatomy' '*die frau ist der welt neher dan der man, und der man ist weiter von ir in der anatomei des ampts halben und das also*'.[62] He focuses on the womb as a microcosm of Creation, while, at the same time, asserting that the womb causes female illnesses. In this sense, the matrix defines the women's bodies and is the cause of female illness.[63] While considering the womb as a microcosm of Creation and as a possible threat to women's health, he creates a philosophical conundrum when he tries to sort out what defines the female body. On the one hand, he regards the womb as a sacred space where the Holy Spirit exists (the womb has a connection to the divine but not to the female body); on the other hand, the womb can harm the woman's body by causing illness (suggesting that there is a connection between the female body and the womb). Lastly, turning to Eve as representative of malevolent forces on the human body (Eve brought illness to humanity), he asserts that there was no illness before women were created.[64] He returns, here, to comparing humans to Adam and Eve; but unlike other works he attempts to define the physicality of the female body.

The *Opus Paramirum* suggests the development of Paracelsus's understanding of the human body as male and female based on his interpretation of Adam and Eve. His emphasis on the visible as a marker of difference and as divine guidance for physicians leads him to focus his attention more on the physical signs visible in the body and in nature. His scrutiny of female physicality differs sharply from the other works considered in this book. Recalling how he previously struggled to define the relationship between a woman and womb, he seems to have reached a critical turning point when he allocated certain forces to nature and then attempted to define nature as itself a womb. In this way the tract is an important crossroads between his theological, medical and alchemical interests. The medieval alchemical tradition regarded the world as comprised of wombs in which nature was itself a womb and chemical transformations were viewed as rebirths. Paracelsus builds upon this tradition when he implores his readers to understand the physical bodies as having varying degrees of relationship to the natural realm.

Paracelsus teaches his followers to delineate the special medical needs of men and women in his tract, *Liber de matricis*. In this way, he further defines his kind of medical practice by finding groundwork in alchemical philosophy to underscore his new and different approach to medicine and to understanding the body. He continues to emphasize the biblical nature of his medical mission using metaphors of the Fall. Lastly, this tract shows how he deviated from his own previous assumptions about the body and clarified the duality of the human body itself.

Von der massa und materia aus den der mensch gemacht ist worden (c. 1536)[65]

Paracelsus's thoughts on the connection of human bodies to the natural world continues in *Von der massa und material aus den der mensch gemacht ist worden*, but he expands upon the chemical composition of the natural body using alchemical paradigms and language.[66] In this tract he is less concerned with the distinction between male and female bodies and instead focuses on embodiment: humans are both natural and spiritual. In this tract, he considers the human body a fifth element that contains parts of the four other worldly elements:

> Humans are made from clay (material) and it follows that humans are the smallest world, that is the microcosm. Because humans are the whole world, in that he is pulled from all the stars, the planets and from the firmament, from the earth and all the elements, he is the fifth being. For the four elements make the whole world and humans are made from them.
>
> *Aus der massa ist der mensch gemacht, aus dem nun folgt, das der Mensch ist die kleine Welt, das ist micrososmos. Aus der ursachen das er die ganze welt ist, in dem das er ist ein auszug aus allen sternen, aus allen planeten, aus dem ganzen firmament, aus der erden und allen elementen, und ist das fünfte Wesen. Dan die vier elementen seind die ganze weldt und der mensch is aus inen gemacht.*[67]

The term 'masse' is commonly found in alchemical tracts of the medieval period and one would find similar explanations for the formation of the philosopher's stone.[68] Even though Paracelsus incorporates alchemical language and constructs, he nevertheless continues to consider the relationship of the physical body to the divine. He articulates his understanding of the physical body as composed of elements of the worldly surroundings all the while emphasizing that humans' physical body was not directly created by God.[69] He uses the analogy of Adam and Eve to human physicality in order to define the substance of the physical body. His comparisons to Adam and Eve force him to consider that human physical birth lacks a direct connection to the divine and he repeatedly differentiates between the physical and spiritual aspects of humans. In this case he explains how the two bodies must be nourished differently:

> He is made in God's image, yet not in godly food rather in the elements. And [he] is divided in two parts in the elementary body, that is, blood and flesh, for that reason he's obliged to nourish himself from the same body. And in the spirit, from the spirit of the stars and he is therefore obliged to nourish his spirit from spirit of the stars.

> *ob er gleichwol in götlicher biltnus ist, iedoch aber nicht in götlich speis sonder in den elementischen. Und ist geteilt in zwen teil, in den elementischen leib, das ist in blut und fleisch, dorumb auch gezwungen von demselbigen leib sich zu erneren. Und in dem geist aus dem geist des gestirns drumb er auch gezwungen wird aus dem geist des gestirns seinen geist zu erneren.*[70]

He suggests that health derives from the proper nourishment of the physical body and thereby underscores a nature-based remedy for treating the physical body. In other words, he begins to articulate a chemical approach to medicine that aligns the physical body with the earth and the natural world.

His emphasis on reading the formation of Adam and Eve through the earth distracts him from his previous claim that his approach to medicine always took into account the divine qualities of human birth. Indeed as he argues for understanding the physical body as related to the natural world and he differentiates between the birth of Adam and Eve and that of all other humans, he redefines the process of conception as part of the natural world:

> Recognize that the hand of God made the first humans and he made them in his likeness in blood and flesh, and it became a human. After that the first human and his wife were subject to nature, and never again did the hand of God create a miracle, rather it all happens through nature, and therefore nature commands the creation of humans and it comes from a particular *massa*. The *massa* is called the seed in nature. But it is not man and woman alone that make a human, rather the elements and the spirits of the stars.
>
> *Nun wissenent aber die hant gottes hat den ersten menschen gemacht und wunderbarlich in sein biltnus und in blut und in fleisch, das er ein mensch ist. weiter aber ist der erste mensch und sein frau der natur underworfen worden, von der hant gotes komen und nimmer ein wunderwerk durch die hant gottes sonder durch die natur, und also der natur die menschenwerdung befolen und in ein massa verordnet aus der er komin ist. Welche masse der samen in der natur geheißen wird. Nun wissen auf solches das man und frauen alien den menschen nit mügen machen sonder sie zwei, die elementen und der geist des gestirns: die viere machen einen Menschen.*[71]

His desire to connect the physical body to the natural world leads him to think about the composition of the body as related to the earth, but his writing reveals his unease with distancing the divine from the natural world and he returns to reminding himself that the natural world is part of God's creation. Nevertheless, his theory that the physical body could be healed through elements of the earth leads to a new approach to pharmaceutical medicine and his followers seem to have remembered that message along with his call for recognizing that the care and composition of women differs from men.

The idea of human birth put forth in this tract does indeed suggest that Paracelsus strayed from his initial idea that there was divine involvement in all birth. Furthermore, if one were to use the tract *Von der massa und materia aus den der mensch gemacht ist worden* to understand Paracelsus's thoughts on human conception it would be tempting to believe the myth that he fabricated human life in the laboratory. In *von der materia and masse* Paracelsus dissociated the material of reproduction, the seed, from the divine. However, when read in the context of his many other writings on human birth it is difficult to imagine that he would attempt to concoct human life. Throughout his oeuvre, Paracelsus highlights his knowledge of human birth and devotes a large portion of his writing to the sanctity of human life. While he set out to affirm his place in the field of medicine by displaying his expertise in reproductive health, he ultimately establishes his place in the history of medicine as an expert in understanding the composition of the human body as related to the earthly realm.

The tracts considered in this chapter show how contemplating the relationship of all humans to the biblical figures Adam and Eve helps Paracelsus clarify his thoughts on embodiment. He moves beyond his early understanding of all human conception as divine and begins to reconcile the difference between the special births of Mary, Christ, Adam and Eve, and human births. In these tracts he maintains his stance that one should avoid learning about the body from classical texts and he still emphasizes the importance of reading the book of nature, but he is less critical of the practice of other physicians and is instead reflective. He asks himself questions that obviously vexed him in his earlier works: how much of conception is controlled by God? Do humans have different qualities than Adam and Eve? What is the significance of the womb in human conception and gestation? Are the seeds of reproduction within the divine realm, or is there an aspect of nature outside of the divine? In asking these questions he articulates dualistic properties of humans: spiritual and natural, which he explains as the birth of two bodies. He takes this dualism further as he expounds on the necessity of understanding the physical qualities of male and female bodies as distinct and different.

Paracelsus interprets the relationship of humans to Adam and Eve as analogous to thinking about the body in the context of the natural world, independent of God. However, his understanding of Adam and Eve changes. He moves from postulating that both were created directly by God, to the notion that only Adam's body was divinely willed. Furthermore, in the end Adam alone represents the natural body, opposite Christ's spiritual body. Paracelsus's shift in viewing Adam, rather than Eve, as the more significant creation in the divine plan alters his thinking about reproduction and, indeed, his understanding of the female body. When he aligns the matrix with the original womb that created Adam, he redefines women's bodies as reflecting natural order and as acting as a universe for the small world within her. However, when he writes of Adam as God's chosen human creation he defines the female body as the womb that passively receives the reproductive material from man. Whereas his early works called for equality of men and women in reproduction, these late works recognize two distinct bodies with varying relationships to the divine. Thinking about the origins of Adam and Eve facilitates for Paracelsus a way to understand humans as consisting of dual qualities that are born in different ways. He distinguishes between the natural body as it relates to Adam and Eve and how the birth of a natural body necessitates a new way to understand the body as composed of natural elements. The works in which Paracelsus reflects on the origins of Adam and Eve reveal his understanding of embodiment: that humans consist of a spiritual and physical self. He argues that the physical body should be understood in the context of the natural world and the spiritual body provides a connection to the divine. He implies that the good physician should recognize and respect the dual aspects of humans. Ultimately his stance against interfering in human birth remains because of the sanctity of the embodied person that can only emerge from the female womb. In this way, Paracelsus urged his followers to respect the female body and to recognize the earthly qualities of humans that differ based on their sex (male or female). He succeeded – not necessarily in the field of women's reproductive health as he envisioned, but his followers did focus on the special nature of the female body as the following chapter shall explain.

5 PARACELSUS'S THEORY OF EMBODIMENT IN THE POPULAR PRESS

> 'In Deutschland sehe ich eine neue Schule an Boden gewinnen, deren Urheber Theophrastus Paracelsus ist', Georg Joachim Rheticus an Joachim Camerariuss 1 Februar 1563
>
> 'In Germany I see a new school of thought finding favour, and its source is Paracelsus', Georg Joachim Rheticus to Joachim Camerariuss 1 February 1563

The second half of the sixteenth century brought some of Paracelsus's writing to print.[1] More and more people had access to his works and interest in his approach to the body grew. One of his prominent followers, Leonhard Thurneisser zum Thurn (1531–*c.* 96) helped foster the paracelsian school of thought in his medical self-help book. He combined Paracelsus's teachings on human reproduction with other profitable methods of explaining the body, such as lift-a-flap anatomies and uroscopy charts.[2] Thurneisser's work demonstrates how some of Paracelsus's ideas informed a way to think about male and female bodies in different ways, how the name Paracelsus began to be associated with a particular kind of medical practice, and how Paracelsus's understanding of human birth was popularized in the second half of the sixteenth century. This chapter explores the spread of paracelsian ideas about the body from the late sixteenth century to the early eighteenth century in the works of Leonhard Thurneisser (1531–95/6) and alchemist/physician Chistoph von Hellwig (1663–1721). The works of Thurneisser and Hellwig demonstrate that Paracelsus's theory of conception and his understanding of the body as consisting of bipartite or tripartite qualities persisted. Their writing also suggests that followers in the field of medicine focused on his alchemically-based approach to the physical body and in this way his teachings had a lasting impact on the way people understood sexed bodies.

Thurneisser and Hellwig were popular medical writers of the sixteenth and seventeenth centuries. Examining their widely-sold works offers a new perspective on the influence of Paracelsus. For the most part, previous research tried to situate Paracelsus in academic medicine.[3] By trying to find the legacy of Paracelsus in the academic realm, he was regarded as a failure because his approach to medicine was not integrated into university education.[4] However, as Rheticus noted in the introductory quotation, Paracelsus's writings were indeed finding an audience. I argue that Paracelsus's success was widespread outside the narrow confines of academic medicine in part because he rejected the reliance on classical texts that typified university medical education and instead insisted on an alchemical-based theory of the physical body. Leonard Thurneisser recognized the value of Paracelsus's alternative approach to the body and relied heavily on Paracelsus's teachings in his anatomy book, *Confirmatio Concertationis*, a self-help medical book. This popular self-help book played a pivotal role in introducing Paracelsus' teachings to an increasingly curious and receptive audience.

There is reason to believe that Thurneisser was introduced to the philosophies of Paracelsus at a young age. Having spent his youth in Basel, Thurneisser grew up in a region that had ties to the famous Paracelsus. Indeed, he apprenticed with a physician who may have attended Paracelsus's contentious lecture in the medical school in Basel: Dr. Johann Huber (1507–71). During Thurneisser's apprenticeship, he read some of Paracelsus's manuscripts and documentation proves that he later exchanged Paracelsus's manuscripts with Toxites.[5] And at the same time Thurneisser was apprenticing with Dr. Huber in Basel, Adam von Bodenstein (1528–77), the influential paracelsian physician, was collecting the works of Paracelsus.[6] Bodenstein gave several lectures on paracelsian medicine in Basel until he was expelled in 1564.[7] Thurneisser's birthplace was an area where Paracelsus and his writing were well-known.

In Basel, Thurneisser had connections to paracelsian medical discussions and may have been greatly influenced by this philosophical perspective, but there is much more to the story of how he came to promote Paracelsus's approach to the body: he recognized the marketability of the paracelsian writings that were being sold at book fairs. The financial success of Paracelsus's writings was evident to Thurnhneisser around the time he opened his publishing company. Thurneisser was an early modern entrepreneur who found many ways to increase his wealth, even though in the end, he was ultimately bankrupt.[8] First, there is the legend of his dubious business practices in Basel. He was prosecuted for selling counterfeit gold to a Jewish businessman. When his deceit was revealed through a court proceeding, his citizenship privileges were revoked and he was forced to leave his hometown.[9]

This unfortunate encounter with the law did not deter Thurneisser from recognizing the potential profit of telling the story of his own trickery himself. His infamous dealings in selling counterfeit gold probably were a source of amusement for those familiar with his name, and he used this fame to his own benefit when he wrote an autobiography about his expulsion from Basel. He explains that it was not his fault, but rather he was framed after the Jewish businessman did not want to pay him. His divorce was also misunderstood: he was forced to marry when he was young and the divorce proceedings against him were all wrong. His turbulent relationship with his brother and the fact that his own brother spread word of his deviance was explained when Thurneisser called the brother an untrustworthy liar who wrongly warned people. Finally, the autobiography explains his special connection to Paracelsus: the illness he endured did not respond to medical treatment until he resorted to paracelsian medicine.[10] The autobiography furthered the association of his name with paracelsian medicine, though he made it clear that his life was distinctly different and more interesting than that of Paracelsus. Through his writings and the legends about him, Thurneisser achieved a kind of celebrity unlike his predecessor. Whereas Paracelsus's personal life remains a mystery, except for a brief note about his alleged asexuality and drunkenness in a letter written by his student, Johannes Oporinus (1507–68), Thurneisser's private life was very public.[11] And it was not only through his writing that he achieved notoriety. We know that he was fond of having his portrait drawn and that he included a self-image in most of his publications, so his appearance, too, was recognizable.[12]

Thurneisser made a celebrity of himself, in a way that our contemporary culture is all-too-familiar. Considering Tara Nummedal's work on alchemists's personae in the sixteenth century that posits that deception became synonymous with the practice of alchemy, it must be understood that Thurneisser's public statements about his past reinforced his medical expertise by discrediting the accusations of deception.[13] He provided evidence of his trustworthiness when he acknowledged the many rumours swirling around him in his autobiographical book with the typically long title: *Ein durch Nothgedrungens Ausschreiben Mein: Leonhardt Thurneissers zum Thurn, der herbrotttischen Blutschands-verkeufferey, Falschs und Betrugs: Auch der Mir und meinen Kindern, zu Basel beschehnen Injurien, Gewatldthat, Spolirung und Rechtsversagung halber (My important writing: Leonhardt Thruneisser zum Thurn, the scandal, lies and deceptions against me and my children that happened in Basel, the injuries, the act of violence, the legal decree).* He also plays up his travels and his accomplishments that are meant to lend authority to his writings about healing the body. We do not know if Thurneisser ever had any formal medical education, though he did become known as a doctor. At some point, he gained the attention of the Count of Brandenburg, Johann Georg and was summoned to help the Count's wife recover from an illness. When he succeeded in curing her ailment, Count Georg named Thurneisser court physician. From then on, it is said that Thurneisser assumed the authority of medical doctor.[14]

Thurneisser's reputation as a physician has always been tied to his desire to make money, even though his contribution to the field of urinalysis using distillation led to new approaches to chemical medicine.[15] Previous scholarship overlooked how his business success provides a way to trace the ways in which Paracelsus's ideas continued to resonate in the works of medical writers for at least two hundred years after his death. Thurneisser was first and foremost a business man. When he moved to Berlin he found financial success in selling astrological calendars, horoscopes and beauty secrets. With his amassed wealth, he opened a laboratory and publishing company in Berlin. He used his laboratory to analyze urine samples which were sent to him by wealthy patients across the Holy Roman Empire. Thurneisser cooked the samples to examine the precipitate and prescribe a remedy for their ailments. He built up a significant clientele and sold many of his elixirs for profit, which he used towards his publishing company.[16]

It is through his own publishing company that he produces what he calls, in Greek, *Bebaiosis agonismou*, or 'establishing rivalry' followed by the Latin title *Confirmatio Concertationis* in 1576. The work draws on several popular medical concerns at the time: revealing secrets of nature through paracelsian ideas, lift-a-flap anatomy, and Thurneisser's own uroscopy success. In addition to combining three otherwise separate publishing phenomena, the book provides ample entertainment through imagery and poetry.

Confirmatio Concertationis was not Thurneisser's first book, but seems to have been his most successful. His letters indicate that it was sold at book fairs in Frankfurt and Leipzig. At least three hundred copies were available through these fairs, but additional copies were sold directly to nobility prior to the general distribution in Frankfurt and Leipzig.[17] Thurneisser's two hundred page book asserts that it will explain the very useful art of urinalysis and will also explore all the internal organs of the body while revealing the illnesses specific to each organ.[18] According to Gabriela Spitzer *Confirmatio Concertationis* was, to use a modern term, a bestseller.[19] She attributes the success to the lift-a-flap male and female anatomies, but I shall argue that it was also his paracelsian approach to the body that facilitated a new and provocative way to think about sexed bodies.

The lift-a-flap anatomical display in *Confirmatio Concertationis* was relatively new. Broadside versions of lift-a-flap anatomical images were available in German speaking regions as early as the 1544. Heinrich Vogtherr's single-page prints were the first to use movable parts.[20] Generally speaking, broadsides were usually a single body, male or female, with the flaps either connected before the sale; or the broadside came with instructions for how to assemble the lift-a-flap anatomy after purchase. The inclusion of lift-a-flap anatomical figures in folios may have originated in Thurneisser's book.[21] The detailed illustrations throughout the book carefully depict all internal organs known to Thurneisser. The full body male and female anatomies appear on large fold out pages that show a distillation chart on the left side of the page and on the right the naked body depicted a natural setting. The female figure includes fourteen detailed flaps and the male figure contains twelve flaps. The additional two flaps on the female model show a uterus and a gravid uterus. The distillation chart (see figure 1) depicted with both anatomical models explains how the precipitate of the cooked urine corresponds to distinct aspects of the body and how one can diagnose certain diseases based on the composition of (See figure 1) the urine. The female model and distillation chart contains much more detail than the male version. Throughout the book Thurneisser aimed to distinguish the features of male and female bodies. He places particular emphasis on the difference in lung capacity when he draws the cross-sectional diagrams of the lungs and the uniqueness of the female body to convert blood to milk for nursing.[22] In the diagrams and throughout the book, Thurneisser, like Paracelsus, emphasizes the need to recognize the uniqueness of male and female bodies. The development of his sex-specific uroscopy charts underscores a paracelsian chemical approach to the body. Yet, Thurneisser goes even further to put Paracelsus's theory into practice when he presents the lift-a-flap images.

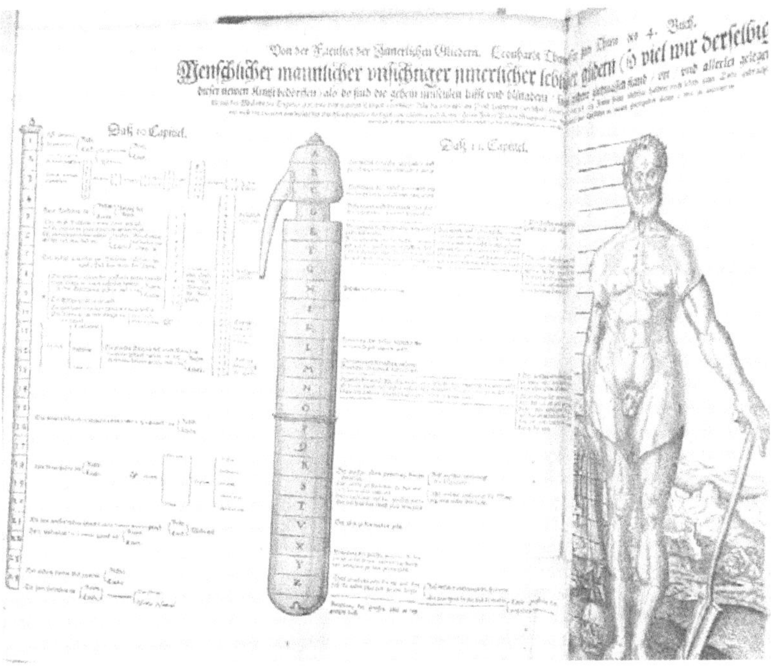

Figure 1. From L. Thurneisser zum Thurn, *Confirmatio conertationis*. Bernard Becker Medical Library, Washington University School of Medicine, St Louis.

Visually, the book has much to offer both educated and uneducated readers and it provided tactile sensation too. It suggests to the reader that, like educated physicians, they can partake in the study of medicine by engaging in mock dissection through the lift-a-flap anatomies and by performing their own distillation tests. Although this visual aspect of the text starkly contrasts with the exclusively verbal presentation of medical ideas in Paracelsus's work, it is nevertheless a very paracelsian approach to the body, for we know how frequently Paracelsus called upon his followers to observe nature, rather than relying on the expertise of classical texts. Here, Thurneisser allows readers to observe the otherwise hidden nature, namely the interior of the human body. In this way, Thurneisser leads his readers on a paracelsian study of the body that also echoes the teachings of Paracelsus explicit in the printed text. All the while he presents the material in an entertaining and engaging way that, judging by the success of the book, invited numerous readers to ponder a paracelsian chemical-based approach to the body.

Thurneisser was likely familiar with the financial success of the pamphlet lift-a-flap anatomies, and of the highly successful reproductions of Veselius's anatomy in the work of Herman Ryff (ca. 1500–48) that were around in the 1540s. Again, this is an instance in which we can assume that Thurneisser observed what was financially successful at book fairs and combined it with other successful writings about the body. Thurneisser did not replicate Ryff's approach to medicine, for Ryff was not a paracelsian. As William Eamon pointed out, 'Unlike Paracelsus, Ryff had no alternative school of medicine to offer, no theoretical axe to grind. Like most popular medical authors he wanted to 'reform' medical practice only in the sense of bringing it in line with traditional medical theory, not replacing the old theory with the new one'.[23] As we know, Paracelsus strongly opposed reliance on classical texts that informed the approaches of Ryff and other more traditional physicians. Paracelsus stood for a new and fresh approach to understanding the body and Thurnheisser publicized this remarkably different medical theory.

With Paracelsus's tracts finding favour in popular press about the same time that Ryff's images were widely sold, Thurneisser took two profitable (though different) approaches to medical knowledge and combined them. He used visual imagery, like Ryff, but in a very paracelsian way. He went to great lengths to demonstrate the differences between the male and female bodies. By this I do not mean that he showed how women had a womb; which was common in the Vesalian anatomies of the time, he attempted to show that the workings of the female body were distinct and separate from the male body.[24] Thurnhiesser's text differs from what would be expected following Laqueur's thesis in *Making Sex*. Laqueur asserted that during the Renaissance the anatomical representation of male and female was more dependent on the cultural politics of representation of the sexes, than on the evidence about organs, ducts or blood vessels witnessed in dissections. His argument that the sixteenth century adhered to a one-sex model of the body that neglected to represent the variance in the female body has been challenged by focusing on the practice of anatomy and in particular skeletal structue. Yet, we must recall that even though Laqueur differentiated between the visual representation of bodies and medical texts about bodies he concluded that both adhered to a one-sex model of the body. Laqueur argued that the discussions of anatomical difference, like those instigated by Paracelsus, related more to a metaphorical understanding of the universe as a womb, than to any kind of physiological difference. The images Laqueur selected for his book demonstrate the conflation of male and female sex organs when they reveal a confusing image of a female body that quite literally look like male sexual organs inverted. The works of Paracelsus and Thurnheisser necessitate a revised understanding of how sixteenth-century medical texts understood the body. The paracelsians did not understand the womb entirely metaphorically as a microcosm of the universe, though this idea does recur in their works. In fact, Thurneisser's book does not at all adhere to such theories of inversion. Like the 'science' that Laqueur cited in the eighteenth century that was used to underscore the weaker qualities of women based on their physiology, Thurnheisser following Paracelsus, insists that the relation of mercury, sulphur and salt differs in men and women. In other words, Thurnheisser suspected that the hidden qualities, the insides of the bodies, revealed sex difference. He theorized the difference in bodily fluids, rather than in the bones, but Thurneisser wanted to stretch the differentiation further by including skeletal diagrams. Thurneisser's images show the slightly wider width of the hip bones in relationship to shoulders on the female than one would expect on a drawing of a male body (See figures 2 and 3). This wider width of the female skeletal structure was a development in anatomical illustration that Laqueur attributed to the eighteenth century. Although Thurneisser's book does suggest some degree of skeletal difference between men and women, he focuses less on the anatomical structures and more on the sex differentiation that occurs on chemical level as evidenced in their urine or the function of their major organs. All in all, he was careful to show the physiological connection to the chemical variations. And to Thurnheisser the workings of the bodies underscored women's inherent weakness.

Thurnheisser's use of anatomical models differed from what was available in print at the time. The lift-a-flap male and female anatomies are clearly labelled to show the connection between organs and the kind of precipitate one would find in urine samples. Letters on the body correspond to the distillation chart. The distillation chart explains the corresponding organ's role in how we understand bodily fluids such as blood and urine. In some cases, but not all, the organ is named on the distillation chart and one can find a longer explanation about it

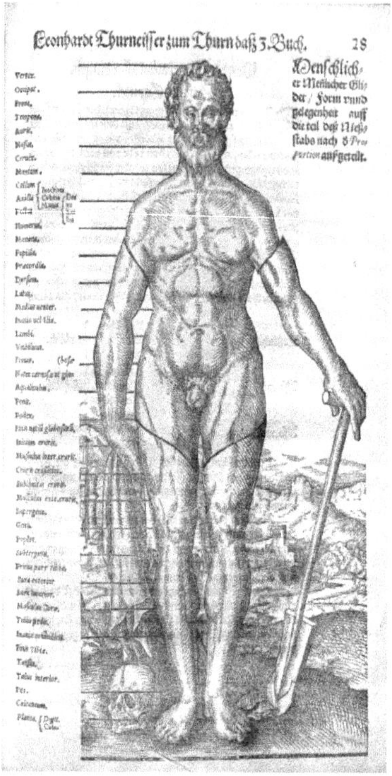

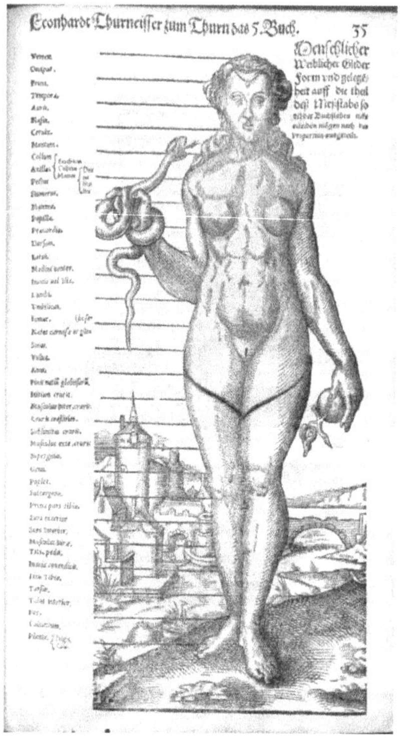

Figures 2 and 3. From L. Thurneisser zum Thurn, *Confirmatio conertationis*. Bernard Becker Medical Library, Washington University School of Medicine, St Louis.

in the following pages of the book. To the right of the drawing, the organs are named and one finds a short explanation about the function of the organ. Very much like Paracelsus's writing that tended to focus on the female body, Thurneisser's lift-a-flap male anatomy figure is less detailed than the female anatomy. The male reproductive organs, specifically the testicles, are drawn so that they show the circulatory structure back to the aorta, but we do not find a drawing that connects the testicles to the penis.[25] By contrast, a Latin anatomy text of about the same time by the French physician Andre Du Laurens demonstrates the connection between the penis and testicles by showing spermatic ducts.[26] In a French vernacular text, produced thirty years before Thurneisser's, Charles Estienne provided a detailed drawing of the male reproductive tract and labeled the figure to show how the sperm moved from testicles through the penis.[27]

Although accurate anatomic illustrations of the male reproductive tract existed in Latin and French vernacular text, the male anatomy provided in *Confirmatio Concertationis* does not include such detail. In both Paracelsus's writing and in Thurneisser's book more care and detail seems to go into describing the female body than the male body. However, like Paracelsus's early works (*Book about Birth*, for example) there is confusion between the process of conception and the physiological pathway for the emergence of male and female seed. In Thurneisser's book the penis is only connected to the bladder. However within the text , Thurneisser explains the necessity of the male seed emerging from the penis in reproduction. Thurneisser neglects to visually show the details of male reproduction, while at the same time emphasizing the qualities of the female reproductive system, which includes gestation and lactation. His disregard for the male body echoes Paracelsus's focus on the female body and women's reproductive function. This could either stem from Paracelsus's attempt to carve out his place in medicine by focusing on women's reproductive health, or from a recognition that most other medical works were modelled on the male body as exemplary. Either way, Thurneisser, like Paracelsus, focuses on understanding the secretive qualities of the female body.

Similar to Paracelsus's association of the female body with Eve in his later works, Thurneisser depicts female anatomy as Eve: the female body holds a snake and an apple (see figure 3).[28] Of course, Ryff also depicted the female as Eve, but Thurneisser draws connection between the Eve-like anatomy and the distinctly female chemical processes within her body. For example a comparison of the urscopy charts reveals different measurable values of male and female urine. The charts for reading male and female urine are drawn the same way, and labelled the same way but the female model has more information in the D through G zone of the chart, the part of the chart that discusses the thorax and lower abdominal area (figure 1 is an example of the male uroscopy chart).[29] The female chart is much more detailed than the male chart because it attempts to explain the special ways the female body works to nourish a growing child. On both the male and female anatomy charts, the beginning of the pubic arteries are located in the same place, but the female version explains how the blood and moisture from her body are transformed to nourish offspring.[30] Despite the lack of explanation for the process of reproduction, Thurneisser carefully distinguishes between the function of female body and the male body explaining for each internal organ whether or not one recognizes a difference in the function and/or form of the female body.[31] In the cases where the same organ exists in both male and female bodies, he asserts that the organs of the female body differ more in quality than materially; the female organs are usually weaker than the male organs.[32]

Following Paracelsus's contempt for classical authors, Thurneisser takes a stab at the old-fashioned physicians who rely on classical texts. In essence he points out the lack of understanding the ancients had for the female body. He questions why texts that rely on classical texts to understand the female body neglected to name the female genitals.[33] Indeed, Thurneisser remarks with dismay that the ancients, like Aristotle, did not even name the female parts.[34] He even questions why the female reproductive organs would go unnamed and remarks how amazing (*wunderbarlich*) the uterus is. He calls the vagina the 'a womanly little part' (weibliche gliedlein). His linguistic construction is interesting because the German suffix –lein suggests both something small and is used as a term of endearment. By naming the female reproductive tract he emphasizes how his work differs from the traditional approach to the body and he echoes the sentiments of Paracelsus that regard the female body as a microcosm of the universe, a little world.

The male model reveals less attention to naming the parts than the female figure. Thurneisser devotes a page to describing the uniqueness of the female body and highlights the secretive nature of it. Within this segment, he also promotes a paracelsian chemical understanding of the body when he argues that women have more mercury in their bodies, while men have more sulphur in their bodies:

> The main thing is that the outward organs seem to be a version of man's, but the inside organs are more than what a man has, so there is more of a spiritual and secretive nature to the inside of women, so I say more of the difference comes from the quality of the organ than from the substance of the organ, so each body will have its own measurable difference in the distillation glass. For because of a secret power, the spirit in women is more evident, but weaker, and compared to the man who is stronger, the art of medicine must be altered for each.
>
> *Ursach das ausserthalb eines manglenden/ und wenig Innnerlicher gliedern mer/ dan am man an diesem Weibsbildt gespiert werden/so ist doch solches mehr von Innerlicher Geistlicher und heimlicher ursachen/ also so ich sagte/ mehr von der Qualitet/ den von der Materien/ oder Substanz warden beschen/ also das auff jeden Leib ein besunderer Messtab Distillerglas und bildtnus geornet. Dann weil ein heimliche krafft/ ode in erhaltender Geist im weib vil wircklicher/ aber doch schwecher/ und dargege im Man stercker und vermöglicher ist/ werden auch die künst geendert*[35]

He draws from this analysis an explanation for women's weaker mentality and explains that we must take into consideration the difference between the body of a male or a female, an old person or a young person when making diagnoses.[36] Thurneisser begins the section by reminding readers that the brain, heart, liver, stomach and kidneys are weaker in women. He then goes on to explain the remarkable nature of the uniquely female organ: the womb. He describes the ability of the womb to draw from other organs in the body, namely, the brain, the liver and the heart. The description of the workings of the womb corresponds to Paracelsus's early theory of conception and gestation, as described in chapter two. Like Paracelsus, he explains that after seven days, the necessary ingredients are pulled from the organs to create life.[37] In this way, Thurneisser introduces his audience to a paracelsian approach to understanding the process of conception. He adds another dimension to understanding the difference of the female body when he visually represents the location of the womb.

Alongside this narrative, he shows small lift-a-flap illustrations of the womb and bladder. Other anatomical representations of the time depict the vagina as an inverted penis (so that it is sometimes difficult to distinguish the male from the female body unless labelled); no visual correlation between vagina and penis exists in Thurneisser's work. The male and female reproductive organs are distinctly different. Both visually and in the narrative, Thurneisser reveals the female body to have agency in conception. Thurneisser carefully delineates the unique features of the female anatomy. Within this same section, Thurneisser uses small lift-a-flap drawings of the bladder and uterus to show the physical proximity of the organs within the female body. The multiple layer diagrams represent the physical difference between the positioning of the bladder in the female to that of the male.

Even though Paracelsus did not use visual diagrams, his written word is transformed in the works of Thurneisser to convey a sex-specific understanding of male and female bodies. In the male model a straight line can be drawn from the armpit to the hip. In the female model, the hip bones extend past the shoulder and exemplify the wider pelvis of a female (see figures 4 and 5).[38] Recalling that Thurneisser subtitled the book, *establishing rivalry (Bebaiosis agonismou)*, the anatomical representations provide a contrasting approach to male and female bodies than those provided by Ryff. In addition, the paracelsian ideas on human reproduction build upon this contrary approach to the body. Therefore, Thurneisser's attempt to establish rivalry using a paracelsian approach to understanding the female body ultimately leads to an explanation of the different physicality of men and women. He is able to visually represent the uniqueness of their bodies in a way that Paracelsus could not in the written word. Although the focus on visualizing the female and male bodies suggests that Thurneisser was especially concerned with the physical aspects of male and female bodies, he retains the spiritual connection between the body and mind that Paracelsus emphasized throughout his writing.

Thurneisser's reflection on the spiritual qualities of human birth are evident in the first chapter of the book in which he declares that he will explain the creation and union of human bodies, the work of doctors, the origins of disease and the origins of natural beginnings of all things.[39] This chapter bears the most striking similarity to the writings of Paracelsus on the topics of birth and embodiment. Thurneisser begins the book by reminding the readers of the holy trinity: Father, Son and Holy Spirit. He asserts that we must keep this trinity in mind

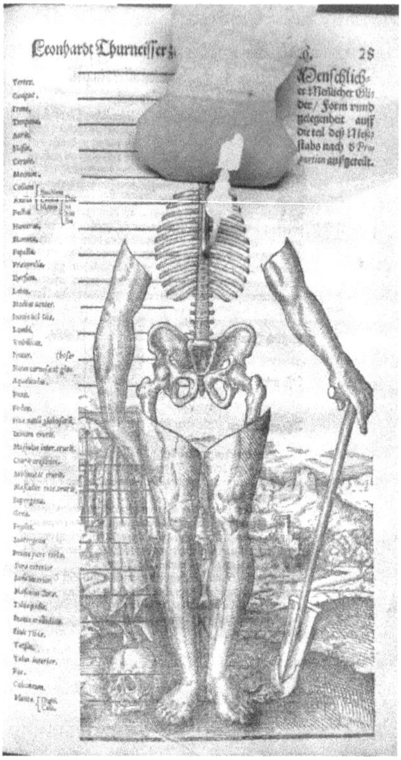 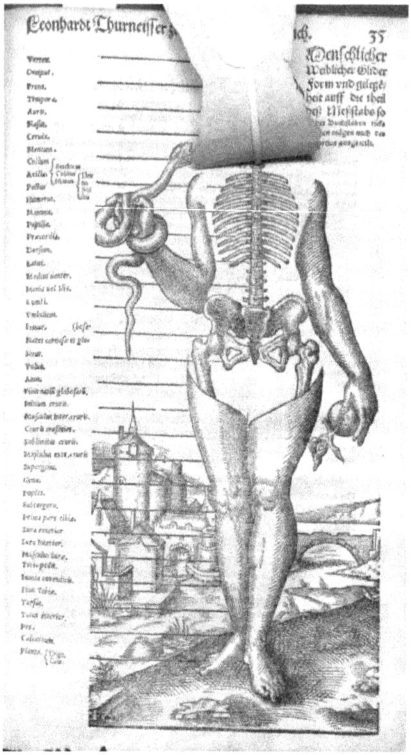

Figures 4 and 5. From L. Thurneisser zum Thurn, *Confirmatio conertationis*. Bernard Becker Medical Library, Washington University School of Medicine, St Louis.

when considering the human, because the human body explains the workings of the entire universe.[40] The marginal gloss that guides the reader through the first segment declares the similarity between God and humans, 'Human composition is the same as God. Humans consist of three bodies' ('*Mensch an Form/ und gestalt Gott gleich. Mensch aus dreyen Cörpern versamlet*').[41] He goes on to explain the three bodies that were united by God in the human body: the essential elements of the heavens, the spiritual or elemental quality, and the natural body.[42] Thurneisser's conception of the embodiment of humans as a tripartite model echoes Paracelsus's later works.[43] He builds upon this notion to explain the mental capacity of humans as a reflection of this tripartite nature; specifically he views the soul as analogous to the heavenly qualities, the spirit as analogous to the elemental properties, and the body analogous to all natural qualities within the human. In this way, he conveys a distinction between physical body and the cognitive elements of human capability. The body physically represents all of nature. Throughout the section, Thurneisser repeatedly breaks each aspect of the human body and mind into three distinct categories to represent the ever present comparison between humans and the holy trinity; he likewise develops a tripartite notion of all aspects of humans and natural phenomena to the extent that he sees three qualities in all the elements and minerals. For example, he writes about human thinking as tripartite: the good, the confused, and the bad/evil.[44] Indeed, Thurneisser emphasizes the concept of a tripartite composition of all things in the marginal gloss: 'Every Substance originates from three beginnings, everything has three substances within it/ every substance originates from three elements', ('Jede Substanz ensteht aus drey anfengen. Jedes ding hat in sich drey Substanentz/ ein jede Substantz entstet aus drey Elementen. Element entsteht aufs den Qualiteten').[45]

In addition to drawing heavily on Paracelsus's tripartite understanding of humans, Thurneisser uses terminology found in many of the texts considered in the previous chapters. He invokes Paracelsus's phrase 'the light of nature' in describing the heavenly qualities of humans calling it the 'light of the holy Spirit' *('licht des heilgen Geistes')* and later goes on to use the exact phrase 'light of nature' *('licht der Natur')* to describe the origins of human intellect.[46] Readers familiar with Paracelsus's works would certainly recognize this hallmark phrase.

It seems that Thurneisser strove to emulate Paracelsus in many ways. He did not stop with linguistic constructions, but rather went on to reflect on what, at the time, was considered paracelsian.[47] Take, for example, the widespread belief that Paracelsus had found a magical stone to restore life to some things. Thurneisser considers this attribution and chooses to follow Paracelsus's own writing, rather than the rumour of Paracelsus's magical stone. He writes of the inability to bring humans back to life, but remarks that 'Humunculos, Scorpion, snakes and other animals, like the Basilisk, can be rejuvenated'.[48] He may have encountered this theory in a Latin text with some German fragments that was widely attributed to Paracelsus, *De Longa Vita*.[49] Reflecting these theories, Thurneisser writes: 'The generation of homanculi occurs through the process of digestion inside a horse and it is like the same from which it comes, in flesh and blood-with more or less limbs' ('*Dan generatio humanculi sit durch grosse digestion im ventre equino gemacht, und sich dem gleich, von dem es komt im fleisch und blut der merern und wenigern glider*'). The idea that the homunculus was generated in and borne from the stomach of a horse is a sentiment that occurs elsewhere in writing attributed to Paracelsus, namely in *Liber de Humonculus*.[50] The idea that the basilisk and homunculus were beings that did not follow the limitations of human bodies was central to some of Paracelsus works on minerals and elements. These texts circulated in printed form at book fairs in the 1570s, about the same time that Thurneisser prepared and sold his book. The text, *De Long Vita* was prepared for publication by Adam Bodenstein in the 1560s in Basel.[51] It is highly probable that Thurneisser was familiar with Paracelsus's tract and very likely that he recognized the monetary success of Paracelsus's texts at the book fairs. With this in mind, we can consider *Confirmatio Concertationis* an attempt to capitalize on the business success of Paracelsus's works by emulating discussions of the unusual nature of the Homunculus and Basilisk.

The topic of human birth, in particular, closely follows the teachings of Paracelsus. Regarding healthy human birth, Thurneisser agrees with Paracelsus's theory of the process of conception. Like Paracelsus, Thurneisser regards the womb as a microcosm of the universe and he explains the formation of a baby through the process of pulling qualities of all parts of the body into the womb:

> For what is more miraculous than this/ that from one good and well-cooked bloom from the most precious and most faraway parts of the whole body...and from the separation of a likeness in its own appearance grows. In over 46 days it will be life, in 90 days movement can be felt. Yes, and in the same way that the fruit grow from so many different organs inside and outside the body such as legs/ skin/ hair/meat/ blood/ and wet substances, it is miraculous and difficult for humans to understand that the human is pieced together in this way and then turns its face toward the mother's back.[52]

Thurneisser's account of the organization of substances to form the foetus is similar to Paracelsus's idea of the *liquore vitae* as described in his *Book about Birth* (see Chapter 2). Recalling how Paracelsus suggested that an image from the mind is projected like a shadow on a wall, which forms the *liquore vitae* and influences the appearance of the offspring, here we see a similar logic in the above passage that explains how substances from the brain, liver and heart are transformed to make the offspring. This explanation corresponds to Paracelsus's theory of how family traits are passed on. Thurneisser explains the process in terms of the organs making an image that transfers to the offspring. It is interesting, and noteworthy, that both subscribe to an idea of reproduction that replicates the idea of printing. In other words, they imagine the process of passing on traits or the process of making the body of the offspring as a sort of woodcut that once inked can be pressed upon the page and form something similar to the original example. Of course, the success of the printing press was probably in the back of both of their minds and they explained how offspring resembled their parents. For Paracelsus and Thurneisser, the process of conception was like a virtual printing press though they differed in the specific organs that contributed to the process.[53] Paracelsus focused primarily on the brain, while Thurneisser included the brain and heart as organs that produced the image necessary for reproduction. Although there are minor differences in how the process occurs, *Confirmatio Concertationis* demonstrates that Thurneisser recognized that Paracelsus's approach to conception was one that people wanted to read as indicated by the relatively large number of copies sold.

Thurneisser did a great deal to make sure that the book had something for everyone. He provided many images that were visually appealing as well as several poems that presumably could be read aloud. All these intriguing aspects of the book introduced, or reinforced for, the reader a paracelsian understanding of the body. Thurneisser also used the various aspects of the book to highlight his authority in the field of medicine. For example, he drew on the popularity of the German poet, Sebastian Brant (1457–1521) when he uses Brant's poem that tells the story of an illness suffered by the beautiful and noble Anna, wife of Ludwig Sturms of Strassburg in 1496. Thurneisser responds to the content of the poem that explains how her malady was so confusing that educated doctors could not muster a diagnosis. He argues that he has a better understanding of the condition and shall explain it. A rough translation of the poem reads:

> You masters of medicine have shown me something that I will now report, something I have never before written about. The story is both curious and noteworthy: The wife of Ludwig Sturms, a fine woman with a beautiful body, suffered with this illness for several years in Strassburg. She cried in agony when loud sounds emerged from her body. The sounds were so loud that they could be heard from twenty paces back. This embarrassed her so much that she was forced to leave church in shame. She sought the advice of many good doctors. [54]

The poem goes on to explain that the physician Georg attributed the problem to a humoral imbalance, a point that Thurneisser wishes to reconsider. Thurneisser begins his analysis of the case by reminding the reader of Anna's remarkable beauty and then goes on to explain that it was not a humoral imbalance (he rejects the classical approach to the body), but rather a problem with the size of her lungs. Thurneisser explains that because she had small lungs she did not get enough air into her body. Her heart, he argues, was large and thus because of the disproportionate relationship of her lungs to her heart, the excess blood had to emerge from her mouth. However, the size of her organs did not explain why worms crawled out of her mouth too. He concludes that the worms were sulphuric in composition and lived in her body because of her lack of mercury. On the one hand, Thurneisser attributes the problem to Anna's physiology and on the other he invokes the Paracelsian *tria prima* (mercury, sulfur and salt) and blames it on an imbalance of the three important substances in her body. Aside from the brief, and rather difficult to follow diagnosis, Thurneisser's response to the poetic introduction to the malady speaks to his overall engaging tone in the book. He seems to entice his readers with entertaining poetry, which both elucidates his points and serves as an approachable means to studying the body. Thurneisser, like the poem by Brant, emphasizes the beauty of Anna's body. In this way it sexualizes the process of studying the illness and indeed the female body. He invites the readers to partake in the pleasure of imagining the female body in this passage and in other sections of the book he provides visual representations of male and female bodies. The female body seems to be of special interest throughout the book. The poem by Brant followed by Thurneisser's diagnosis is not followed with a similar discussion of an illness suffered by a man, nor do any illnesses usually associated with men (like gout) receive attention in the book. Medical tracts by Paracelsus and other important physicians of the sixteenth century usually address men's illnesses in more detail. Furthermore, the use of the poem to introduce a discussion of a beautiful female body sharply contrasts with the straightforward almost clinical description of the male body. Clearly, he sought a marketing edge on other medical books of the time when he invited the reader to engage in imagining the beauty and suffering of a noble woman and he also highlighted the special qualities of the female body.

Thurneisser's entertaining book shows how a two-sex model of the body emerged from paracelsian thinking about conception and gestation. The comparisons to the holy mother of Christ led Paracelsus to contemplate both the spiritual underpinnings of human nature and the special ability of the female body to nurture a developing body and give birth. He began asking what needed to be different about the female body to make it possible to bear children. Though he saw parallels in the contribution of seed to conception, as did Thurneisser, Paracelsus pondered the special properties of the womb. Thurneisser's text speaks to an understanding of an embodied gender when he writes of the weakness of female organs and the weaker quality of the mind. One sees in Thurneisser's work a relational hierarchy based on physiological difference that Laqueur argued originated in the eighteenth century. This begs the question as to why paracelsian physicians' writing about the body has been overlooked. Is it their lack of authority because they did not hold university positions? Is their contribution to the development of modern medicine neglected because they were not necessarily university educated? Perhaps it is a little bit of both, but most importantly in this analysis is how adherence to a paracelsian approach to medicine led them to reflect on the physical qualities of the natural body, that to Thurneisser was sex-specific.

The sex-specific understanding of embodiment led Thurneisser to make, albeit brief, assumptions about gendered life. When he wrote about the weaker lungs of women, he implied a relationship to men as inferior, or at least not as physically strong. This differs from the idea that the female was an inverted or less perfect male. In terms of her reproductive function, Thurneisser did not always correlate the female body with the male, but rather viewed it as different. Thurneisser's book suggests a strong insistence on a two-sex model that was based on difference of internal organs and the difference in chemical composition. It also shows us how some of Paracelsus's ideas were spread throughout German-speaking regions toward the end of the sixteenth century. There is reason to suspect that Thurneisser, through is publishing success, played a major role in keeping Paracelsus's notion of embodiment and his approach to medicine and the body under consideration for another century after Paracelsus's death.

The works of Christoph von Hellwig (1663–1721) demonstrate a continuation of the paracelsian tradition until well into the eighteenth century, but Hellwig also played a role in distorting paracelsian teachings. In many ways, Hellwig was more of a follower of Thurneisser than a student of Paracelsus. While Thurneisser made every effort to elevate Paracelsus's approach to medicine, Hellwig relied on Paracelsus's chemical teachings and eschewed spiritual meaning within the body. Nevertheless, Hellwig maintains the chemical distinction between male and female bodies. Christoph von Hellwig is generally regarded as a paracelsian physician; but the attribution only explains the chemical aspect of his approach to medicine.

Christoph von Hellwig studied philosophy and later medicine in Jena, then he travelled the world with his brother Otto, who was a professor in Heidelberg. He completed his medical education in Erfurt and in 1688 became a physician for the city of Weissenfels.[55] Hellwig seems to have followed Thurneisser's lead in producing medical texts for everyday use. A prolific writer, Hellwig wrote and edited more than forty medical and pharmaceutical books, many of which were dictionaries, household guides and reports of unusual cases.[56] Like Paracelsus, Hellwig explained his decision to write in German because it was more accessible than works in Latin.[57] His choice of the vernacular proved successful: his works were reprinted several times even after his death. Indeed, his notoriety caused some publishers to post-date the reprinted works so that it appeared that the great physician was alive as much as thirty years after his death.[58] Like Thurneisser, he produced a work on uroscopy for general use that was reproduced as many as four times. The fourth edition was published after his death in 1748.[59]

Hellwig's urinalysis is more detailed that Thurneisser because it provides diagnoses based on the colour and quality of the urine. Much like Thurneisser, Hellwig compares and contrasts the urine of men and of women.[60] In this way, Hellwig was following in the paracelsian tradition of understanding the sex-specific qualities of the body based on fluids that Thurneisser popularized. According to Hellwig, women's urine is thicker and has whiter sediment than men's urine. He then goes on to discuss how the urine can reveal illness. This section is divided into 'Special rules for the urine of men', 'Special rules for the urine of women who are not pregnant', 'About the urine of pregnant women', 'Special rules for the urine of young boys and girls', 'Justice from urine based on chemistry', and 'How to measure urine'. Generally speaking, Hellwig argued that one could make a diagnosis based on the colour of the urine. Highlighting the paracelsain insistence of the importance of observation, Hellwig also explained that one could ascertain a young woman's virginity by the quality of her urine-it should appear bright.[61] His section on pregnancy reveals the contemporary dispute about whether or not one could determine pregnancy based on urinalysis when he acknowledges in the introduction to this section that doctors disagree on whether or not the urine can reveal pregnancy.[62] He sides with those who agree that changes occur in the urine during pregnancy. He explains that from conception to the sixth month the urine appears gold-yellow and very cloudy; however he acknowledges that at first the cloudiness may actually be some of the man's seed left over from the night before.[63] Around the seventh month of pregnancy, the urine will become more clear and watery; just before delivery the urine will have blood in it. In the section on women who are not pregnant, blood in the urine was indicative of menstruation. In this section he also suggests that if one finds a small kernel in the urine, then it indicates pregnancy. In these examples, one can see how Hellwig built upon Thurneisser's success in uroscopy and further correlated chemical processes with the patient's sex. Furthermore, Hellwig understands the process of conception and gestation as a unique chemical process.

Between the publication of Thurneisser's book that narrates and illustrates a paracelsian approach to the body, and the publication of Hellwig's book, we see a decline in the philosophical contemplation of the workings of the body in favour of what might be considered a more scientific approach to the body. Whereas Thurneisser interprets the body for the reader as part of a divine system, Hellwig describes the form and function of each organ and only refers to the divine in his anatomical work in the introduction when he thanks God for the inspiration to write the book.[64] One of Hellwig's later works, published under the pseudonym Valentin Kräutermann, reflects a more paracelsian approach to the body. In his 1737 work on eye care he describes the eye as a microcosm of God's creation:

> The first task completed when almighty God created the world was the creation of light so that all the beauty in the world did not remain hidden in a dark abyss. The eye is the light of the microcosm, a guide to all that God created, all that is good. Through the eye we see the marvellous workings of God: the firmament, the world, and their order and pleasantness.[65]

In this work, unlike the anatomical text, he draws frequent parallels between the body and God's creation. He associates the eye with the microcosm and the ability to observe God's creation. This is in keeping with Paracelsus's notion of the light of nature and his insistence on observing God's creation in order to understand the body and its illnesses. It seems that Hellwig drew from different sources to devise his theories of uroscopy and eye care. For the uroscopy it appears that he relied on the works of Thurneisser and in that instance only followed the chemical teachings of Paracelsus. His work on eye care is reminiscent of another paracelsian: Bartisch (c. 1535–1607). Bartisch's *Opthalmodouleia* was self-published in 1583.[66] In many ways, Bartisch's guide to eye care is comparable to Thurneisser's text: *Confirmatio Concertationis*. Bartisch spends a great deal of time in the text promoting himself and his methods, while relying on a paracelsian approach to the body.[67] Yet, Bartisch focuses on the work of the physician as a spiritual and divinely guided mission, much like Paracelsus's self-definition in his early works.

In dealing with the topic of eye care Hellwig echoes the paracelsain approach to the organ as specified in the text by Bartisch. Indeed, Hellwig tends to follow the format set forth by Bartisch when he lists specific eye problems followed by the remedy. Bartisch had introduced readers to herbals that one should drink to cure or improve eye problems. Hellwig likewise provides guidelines for precise measurements and dosages of mostly herbal remedies.[68] Whereas the chemical approach to medicine in Hellwig's work draws upon the medical philosophy made popular by Paracelsus and his followers (like Bartisch), the approach to the body itself does not distinguish between, or contemplate, a notion of physical body as separate from, or part of, the self, as was evident in the works of Paracelsus and Thurneisser. Still, in his pharmacological writings, Hellwig echoes the ideas of Paracelsus and his follower, Thurneisser, when he carefully delineates the way analyzing the body must recognize the uniqueness of male and female bodies. Hellwig insists that medicines must be adjusted according to the patient's sex. He understands pregnancy as a chemical process, revealing both a connection to the paracelsian tradition and the influence of the contemporary advancement in the field that was beginning to be known as science.

Hellwig's writing represents a turning point in medical thinking. His interest in finding an elixir, or philosopher's stone that will cure everything, has led scholars to place him in the pre-enlightenment era, but we must keep in mind the transition in thinking about the body in some of Hellwig's works.[69] The body is not understood by reading the stars, nor by contemplating it as being created through a mixing of substances from the mind. Conception is simply a chemical process for Hellwig. Only the eyes retain the microcosmic quality and connection to the divine. Paracelsus and Thurneisser extended the microcosm to the entire body, especially the womb. Bartisch considered the eye a microcosm. While the uroscopy works of Hellwig diverge from the philosophical contemplation of the body that highlighted Paracelsus's oeuvre, some of the core aspects of treating the body with regard to its sex stem from the teachings of Paracelsus. Although the medical handbook that teaches the reader how to interpret urine samples bears only some reference to Paracelsus's teachings, Hellwig did a great deal to promote Paracelsus's philosophy in the early eighteenth century. In 1730, under the pseudonym Valentin Kräutermann, he wrote: *The thuringian Theophrastus*

Paracelsus, Miracle and Herbal Doctor or the Curious and Intelligent Doctor of Magic (*Der Thüringische Theophrastus Paracelsus, Wunder und Kräuterdoctor oder der Curieuse und Vernünftige Zauber Arzt*).[70] In this endeavour Hellwig participated both in the popularization of Paracelsus's philosophical ideas and in the exaggeration of Paracelsus's mythical identity as a mad Alchemist.[71]

Hellwig's book on Paracelsus explained Paracelsus's understanding of nature and attempted to situate Paracelsus's approach to nature within the context of great philosophers who also recognized the importance of reflecting on the meaning of their surroundings. He emphasizes that Paracelsus viewed this as a divine calling. While Hellwig recognized the way that Paracelsus viewed his work as a physician as part of a spiritual healing, Hellwig attributed various remedies to Paracelsus that did not correspond to any of Paracelsus's teachings (such as a love potion).[72] Indeed, in this particular work, concepts attributed to Paracelsus deviate so far from Paracelsus's ideas about conception and gestation, that if one were to base their understanding of Paracelsus on Hellwig's book, it would be a far cry from the philosophical ideas put forth in Paracelsus's own writing. To this end, it shows how Hellwig's publishing successes merely promoted Paracelsus's chemical philosophy. And though Hellwig did continue to understand men and women as chemically different, a key aspect of Paracelsus's theory of embodiment, he simultaneously presented ideas that discredited Paracelsus.[73]

Thurneisser and Hellwig were bestsellers. Their success suggests that a wider audience than just those educated in the field of medicine, might turn to these authors to help themselves heal, or to help them care for their loved ones.[74] Looking back on the way Paracelsus's ideas were spread to a larger audience reveals that his contemplation of spirituality or mental qualities as distinct from the body may have lost some if its influence in medical writings of the eighteenth century, but it emerges again in the writing of early psychoanalysts as they contemplated the relationship of the mind and the self.[75] The concept of the body as a microcosm remained evident in the work of Hellwig, which suggests that some of Paracelsus's teaching remained part of popular medical literature. More than any other concept in Paracelsus's writing, the idea of a two-sexed model of the body that demands different remedies for healing men and women, as well as distinct and different roles in reproduction resonated in the works of Paracelsus's followers. It is important to keep in mind that Paracelsus's vehement distrust for the medical profession and his reliance on divinely created substances as the source for healing led to the popularity of this kind of handbook, as evident in the writings of Thurneisser and Hellwig. Although scholars have long sought to find evidence of Paracelsus's impact on medicine in terms of academic teaching, the works of Thurneisser and Hellwig demonstrate that Paracelsus's greatest influence may have been in the popular realm, in early medical self-help books like the few examined here.

CONCLUSION

Paracelsus's theory of embodiment emerged gradually from his need to assert himself as a unique physician with a new approach to medicine. He keenly recognized the widespread interest in understanding human conception and gestation that permeated religious, medical and legal discourses of his day. People were curious about when and how a person gained a soul, whether physical deformity was indicative of a lurking evil and wondered if the moment of birth was such a precarious time that women should be supervised in the birthing room. Defining and explaining the meaning of the body took on new significance in the sixteenth century as Reformers asserted the direct connection to the divine rather than the mediated spirituality of the Church. At the same time German universities began to offer medical education and the field of medicine was likewise experiencing a revolution of sorts. Authority needed to be established. When the university-educated physicians asserted their dominance based on knowledge of classical texts, midwives, alchemists and other marginalized practitioners sought to highlight their unique skills. Paracelsus was cognizant of the multiple ways in which the body and soul were scrutinized and while he may have marched to the beat of his own drummer, he never ceased to inquire about the relationship between spiritual being and the physical self. He likely never intended to articulate a theory of embodiment. Instead, he began on a rather simplistic crusade to establish his medical authority that drew on already established medieval alchemical traditions and also his new-found freedom to read and interpret the Bible for himself. It was undeniably a frightening endeavour that became even more dangerous to him as he began to question the limits of the divine. Had more of his writing been published during his lifetime, he may have been hanged or tortured like other radical reformers of his day. But Paracelsus was curious and used his writing as a way to work through his conviction that medicine required attention not only to the body but also recognition of spiritual qualities which drew him to explain that all humans embody both spiritual and physical aspects.

He adamantly opposed the theories of classical authority, so cherished by learned physicians, and instead related physical aspects of human existence to

Biblical figures. It was only through long periods of contemplation that he convinced himself of the dissimilarities of the physical body to Christ, Mary, Adam and Eve and recognized the earthly or chemical qualities of the physical body. As he began to imagine birth as analogous to an alchemical transformation, he stumbled upon the idea that the chemical make-up of male and female bodies differed. This revelation was his lasting legacy in the field of medicine. As a follower of Paracelsus, Thurnheisser promoted Paracelsus's alchemical reading of the body in his uroscopy guide and highlighted the need to understand women's medical needs as different from men's. Diagnosing urine samples put into practice the theory that Paracelsus reluctantly put forth. These popular guidebooks advanced the idea that sex had a scientific basis: one could read sex as one interpreted the body's fluid. This was long before the discovery of hormones, or chromosomal difference, but one has to wonder if the emergence of the idea of the physical body as chemically-based ultimately lead researchers to look for more detailed markers of physical difference. Perhaps the assumption that parallels exist between the emergence of science and the idea of sexed-self are not too far-fetched when we consider that Paracelsus was philosophically investigating the idea that the body could be a natural entity, separate from the divine. It was not laboratory science that revealed to Paracelsus the need to understand male and female bodies differently, but his followers turned his theories into practice. His theory of embodiment had future ramifications. As Karl Jung has explained, Paracelsus's theories of self lead to modern-day psychology and this study shows how his contemplation of the body was incorporated into alchemical medicine. Furthermore, this study suggests that Paracelsus's theories of the natural body could have laid some groundwork for thinking about the sexed-body in natural law.

Paracelsus's repeated attempts to define himself as an expert on conception and gestation need to be considered when we think about his place in history. He advocated a new way to think about birth that emphasized the important role of the female body in conception, he drew parallels between the field of medicine and understanding divine order, he called for compassionate medicine that recognized that the body includes another sacred, spiritual, body within it, and he called for the need to recognize the natural aspects of physical bodies. When German tracts on conception and gestation were translated into English in the later half of the sixteenth century they were building upon Paracelsus's work. Future work on theories of human birth needs to consider the important role Paracelsus played in understanding female bodies. He was neither the mad scientist who created human life in a laboratory, nor necessarily the loudmouthed drunk as some prefer to remember him, but rather a major figure who needs to be included if we are to understand past theories of the body and the emergence of chemically-based sex difference.

APPENDIX

What follows is a translation of Paracelsus's tract, *Buch von der Gebärung der Empfindlichen Dingen Durch der Vernunft*. I worked from the text as it appears in Sudhoff's collected works. The page numbers associated with Sudhoff's collection are included in brackets within the translation so that the reader may compare to the original. Paracelsus's writing style is complicated and I have done my best to convey his tone and to represent the length of his complex sentences. When necessary, I have inserted stops to make his argument more accessible. I concluded my translation at the end of the tract known as *the Birth of Sensible Things*. Sudhoff, following Huser's collection, grouped *Pars de Statu humano* along with the tract however it is unclear if they actually belong together.[1]

The Book About the Birth of Sensible[2] Things in the Reasoning of Theophrasti Hohenheimensis Germani[3]

The first book, the introduction of Theophrastus to the book about birth

First we must examine the Scripture and many of the books of philosophy and those which name earthly gods, which in their writing could contradict what we will discuss in the following text. Even though some of these theories might make us laugh or might provoke us to make fun of them, it nevertheless behooves us to take these things into consideration and respond to them. First we are moved to begin a discussion against our forefathers of philosophy, who have given much effort to write about the birth of humans. Even though the light of nature[4] meant nothing to them, we sense that they also gleaned nothing from it; for it was not like our philosophy which comes from the light of reasoning, rather it derived only from the cunning of their senses. But our task is to respond to them, because they only wrote from their knowledge, not from experience and observation. The experience and hands-on knowledge alone should guide writing and not contemplation and thought. Aristotle, who sharply fantasized, did not leave much work about birth. But in nature a self-taught man, with no knowledge of natural philosophy, but who is astute, could go and observe nature and describe it with rational words, with crafty sentences and witty sayings. [p.

244] While in nature his feet lead the way to writing. Who wouldn't want to experience such a bright and pure light in his nature? The others, his teachers and his students, have wasted such knowledge as this, they practiced reasoning and argumentation, but in the seed of nature a stamen and a cut flower...[5] We pay attention to the earth. The human body with eternal salvation is the noblest. We recognize nature and use her as a guidepost when we philosophize and speak wholehearted...[6] Avicenna (used in the name of all physicians) proposes with his current and former assistants, a very well described teaching about human seed and birth. And as the same have practiced their medicine in order to discover the human formation, their philosophy gives the impression of physicians practice, who work in this way, not from philosophy; for philosophy is a life-giver to good doctors. But they adhere to it, as we say in German, the starting point, not the beginning. And things that are not material, they take it as if it were. Therefore as the end is, therefore the beginning, and all cause must be nature. But why do we want to concern ourselves with the meandering of a man who was educated by nature, not through philosophy? So that we can show you reason, things that work in congruence, death and life. You need to understand astronomy, a field that reckons with objects and rules which lay before us.[7] [p. 245] We are of the opinion that paying attention to small things is useful. When your writing comes to mind about how the planets and stars (following their very fathomable motions) should make and form children, as many have concluded, as many say, they create their nature, give them wisdom, and virtue. But what do we have to tell you that it is not this way? We do not want to make you think that the stars and planets have no influence. The assessment will come in the following chapters; for you will not be content with an introduction, you will only find what you want in the text. We have not drawn on Aristotle. We want to excuse ourselves a bit for this introduction; the text will complete our argument.

The other introduction

Before we lead you into our text about the birth of humans, it is important to have a model which our writing and argument will follow. The main thing is to name the basis of what we see. There are two elements to human birth that we do not find in others, the first is the sensibility (*empfindlichkeit*), that is man and woman. From the pairing of these two comes the birth of a child, who is similar in appearance and shape to his parents, from whom he is born. And we will let it stand in this introduction and at another time we will tell of other births, such as the monoculus, cyclopse and such things. [p. 246] Such varied-shaped beings appear for the most part as something to consider philosophically, that not all humans (that is, not all forms) come from one father, rather they can come from more than one father. We, two-footed beings, come from a special father, the cyclopses come from a special father, and the monoculi also come from a unique father. It should seem funny to us that such people seldom form from seed (*samen*) or become this way from what their mothers see. These ideas are perhaps just grumblings which will briefly be described: philosophy tells us that a mother can change the form of her child by what she sees, otherwise it will be like her or like the father. We will leave these ideas as is for now in our introduction with the opinion that something is different in the formation of the birth of an *ietliche*.[8] It is against nature that a cyclopse and his wife would bear a *monoculum* and a *monoculus* and his wife would bear a *cyclopem*. The same goes for other beings, known and unknown to us. It would be easier to prove that the wild people could trace their origins to two-footed people and that they were born from the same father, rather than to follow the argument that we put forth. And although the wild people are given no reason, rather they are robbed of it, much like animals, this is another sign that they do not come from humans, rather from a different father. Since the eternal (*das ewige*) is present in these differently formed beings, we call them humans. [p. 247] All humans come from one father and from one God, who alone gives them a soul (*die sel*). The very same God who gives and has given the eternal, who philosophers recognize, would not have let so many forms of beings come from one father, rather he would have let these others come from many different fathers; and he would not go against nature, but rather would create an order in nature with much variance. However, this introduction will not explain the father of the first two-footed creature, it would not be philosophical to do so, because it would appear comical. For the main thing is that we understand the first father from whom all children were born, so we must remain still and observe, for what kind of philosophy would tell us how this very being was born? For God created him.

So that we may conclude and turn our attention away from the writings and letters concerning his creation, we make note that such things are hidden to all of us. And if something of this material were presented as if we had knowledge of it, it would appear comical, or it would be an example that no one understands.

The Third Introduction

So let us turn our attention away from that speech and examine our task, that is to tell about the god-ordered origins of children. We use the two-footed as an example, man and woman. It takes two to make a child and not one or three. But here, in this introduction we let this argument rest. Just like those among us bear two-footed, it is the same in other sects – all formed from their own. However we must recognize that there are many different sects and many different forms: so among us the forms are different but our nature is the same. For this reason when we understand our birth we understand the birth of all the others too. For there is only one nature, but it is hidden in many different forms. It is all within humans and not on his body. This philosophy is based on the microcosm, which is within humans and not in the appearance, not before our eyes. We cannot touch it. But the reason we take the two-footed as an example is not because we are of this kind, or because we are of this group, or because we are familiar with them, or because this group possesses this half of the world, no – not because of that. But it is because Christ was part of our sect, he took our form. He, who had the choice to be any kind of being, became one of us without natural seed (*samen*), we notice how the other sects are incomplete in their forms, they have something sick or ridiculous in their shape, the two-footed beings do not have this and we argue that Christ's sect was well-crafted, faultless, whereas the other sects were of weaker health, therefore incomplete in their form. Although the faces of the monoculi and the cyclopses do not hinder their ability to run, they are as complete as they need to be, however a difference could still be found in that they were created from God as part of the nobility of the two-footed beings.[9] [p. 249] For the internal nature, and the nature that is life, is similar in all beings, not weaker, nor stronger, just different. It is difficult to avoid considering why we are rational and have the light of nature within us. But philosophy should not bring us grief, rather you must bear with us as we consider our origins, a deformed being is just a mistake in the creation. Therefore all that is in proper arrangement in the creation becomes a two-footed creature, and that which is disorganized becomes deformed, as the body indicates. For in the same way that there are many natural influences, the deformity being against nature, God wants to have deformity in humans and leave some things incomplete. Therefore we consider the body of a deformed person to have been created by God, but they lack the eternal (*ewigen*). That is why you will notice that such beings give birth to similar children with attributes deformed in nature, because it is in their nature to pass on the deformity. That is why the way it is in the Creation continues in all creation. But the things that nature makes deformed will be unable to give birth, but if is in entity in Creation , nature will allow it to give birth to something like themselves.

The Fourth Introduction

So that we do not digress too far or extend the introduction too much, we want to conclude the introduction after we put forth the the complete birth of all humans, leaving no kind out. We therefore establish that we want to describe all of their origins which have the eternal (ewige) within them. [p. 250] And we do not want to bring up things that we said we would not mention, like the differences in their bodies, we will move away from that topic now. But all the beings that are internally similar we would now like to consider. As an introduction to our book, as mentioned already, we are not explaining all creatures in the world just those in our quarter of the world. Not that the world is three times bigger and than we are aware, rather the part that we inhabit is only a fourth of the world. We recognize that there is a world in the water, in which the same kind of people live, in which the eternal is present and the same in body and nature. Even though we sense and recognize that there is a world in the water, another book would need to be dedicated to their births. Then we would be led to a third book for the people who live in the earth and soon after the conclusion of the third book we would be inclined to write a fourth about the volcanic people, who live in the fourth part of the world and build fire from their elements.

Therefore recognize here that we examine a fourth of the world, which makes a complete world. We should not let philosophy take nor draw us from this task. We do not want to let ourselves diverge from earthly masters of the soul. [p. 251] We will not let your hallucinations and opinions deter us, we recognize your misunderstandings. We want you to approach our text with good sentiments and we hope to enlighten you with the teachings of Christ. Christ gives us our foundation and our belief in Him is the basis of our philosophy. As evidence brought forth with the birth of Christ, as our philosophical experience dictates, we offer this text, our books about the four births of the world. The first book from our world as the *aereis*, the other from nymphs such as waterpeople, the third from dwarves, such as the mountain people, the fourth from the volcanic people. So we conclude with the birth of the whole world, in which the eternal is the soul (*sel*), the highest calling, the greatest piety, the greatest reward, from the saviour to all of us, God and man, Jesus Christ, born from the virgin Mary, who remained so before his birth and remained eternally virginal, remains in the holy trinity for all eternity.

End of the fourth Introduction.

What follows is the text of the first book of Theophrastus. [p. 252]

Tract 1

Chapter 1

To begin we present a discussion of how we undertake our analysis. We separated the birth of all created things in two ways. The one is the creation (*etliche*), that have the seed (*samen*) of their procreation within them, mixed in and grown naturally in the nature that they have, therefore their nature and their seed are inseparable, they are mixed together and cannot be broken apart. Such is the case in irrational things (*unempfindlichen*) such as grass, stones, metals and others. The things that the rational spirit has, from their birth will be explained. The other difference between the *etliche* and the rational beings is that they have no seed coming from their nature, rather they are completely without seed. These are the humans and animals. No man has seed within him, no woman has seed within her, that comes out of their nature or is born within them like the *etlichen*, just like the other animals. But we are not going to discuss the animals any further and instead we focus our attention on the birth of humans. The above should be a guideline for what we will discuss. The conclusion is that all sensitive beings have no seed within them. From this chapter you should understand the fact that humans do have seed that grows within them naturally, nor do they get seed from nature to reproduce and give birth. Nature has no power over them and the humans do not need seed. Therefore it is only nature without seed and without mixing with nature and the seeds of the irrational clearly are within them, just not in humans. [p. 253] So you should recognize that the *entpfind-lichen* beings have a things for themselves which is a mother and no man, and what will be recognized in human is that also a mother alone if it carries seed, but in humans this is not the case.[10] For if one is a man he will have no seed to regenerate, for women it is the same, they have no power over seeds. Therefore they are without seed in open nature, in which no seeded-being is born. On this basis all irrational beings are mothers: the rational definitively not. Therefore the humans are free without inborn seed and do not have the unification like other beings.

Second Chapter

As we already presented, the human has no seed within him that comes from his nature, we want to prove to you with an example, so that you will know, that beings like you do not reproduce from nature, that you will know that you are free from seed and not coerced, you are pure and free in nature from which you have the light of nature; wherever one finds seeds within the nature of the being there is no light of nature, rather it is dead. Here is an example. You know that a human with seeds has no reason and lives irrationally, he is alone and fantasizes. Even though we just mentioned a human with seed and we mentioned before that humans have free nature, we will explain this in the following chapter. In the first text we note that God created humans to be intelligent and keen, so that humans are free within the light of nature. That is why we want to consider all contradictions, that a divine philosophy and one who wishes to be a natural philosopher must live in the light of nature and not in the seed of nature. [p. 254] So that we may explain the whole reason, you must note that God created all things and only placed humans in the light of nature, free and simple. For this reason he took seed from from nature and did not build it into him, because the eternal is within him. For he wants a human not an altered human, like the seed does to humans when it is within them. This we recognize as a sort of conclusion. He created man and woman and gave neither of them seed, so that they were not led away from the light of nature. For the purposes of procreation he gave them free will, if they want to or not, they may give birth and give them seed. And he put the seed in their fantasy, materially and with all his being. If the man wants to, his thoughts (*speculation*) makes within him a desire, the desire makes the seed, as we will explain later. Therefore God placed the seed in his thoughts (*speculation*) and gave the thoughts free will, to make him want or not and not through nature, only throught the object. When a man sees a woman she is the object. For God gave humans reason, so that they know they can desire. A human can want to become complete or not, he can use these resources or not, he can follow his reason or not. For this reason God puts seed in human's thoughts, in which all reason rests and the object ignites. Only when the human desires does he have seed within him, but otherwise there is no seed in him, only in the light of nature.

Chapter 4 [p. 255]

What has been said about men is also true of women. When she sees a man he is her object. Thus follows in her thinking what the man is, and this understanding is God given. So if she is eager or not, if she desires him, she will have many seeds, when she does not there will be no seed, only air. Therefore God placed seeds in the desire, within the human will, whatever he wants can be done. And if the desire is there, it must be present in both, that is in women and men; therefore both their desires make the work out of their wills. Thus their combined desire gives birth to the seed that no human sees or wants to see. Why then do the blind doctors fill so many chapters with discussions of sperm, as if sperm were a seed, and it is no seed? They want to use their skills to help women get pregnant, to make sperm, these things are not possible with their skills. For it is not a material thing, it is only an act of thinking, and to make more sperm is an *egestio humoris* (digestive humour) and not of the blood, and in the places where we are told about sperm, and not as from seed, rather from a superstition of nature in which there is no seed....Though many foolish people have written about it, they all write about it the same way. It is foolishness that anatomists say that sperm is the seed and they speak about the nature which bears this as human and should really be from God. So that their piety is not scrutinized because of this desire, they say from their wobbly stools that humans should do as humans are. They live by this saying of Aristotle, as he says that nature must have its ways of revealing itself. We credit Aristotle with this thought, but not the idea that seed comes from nature, like the satanic whores who excuse themselves of their evil ways. [p. 256] We are fascinated by the god Tibullo,[11] and whether he wanted to be purified by the light of nature when he wrote of lust...Maybe he was never uncomfortable or he slep and let the muses dance. Why do we worry about the gods that doctors mention? Because something is wrong with their philosophy, yet our text is incomplete and we want to turn our attention to the birth of humans as we said we would in the beginning.

Chapter 5

As we have shown, the seed lies in the speculation and desire is a mother of seeds, and the human is incomplete and does not have a seed for himself, rather for another, this is, he is only a half seed. We want to show why God made humans with only half a seed. Therefore we propose that such an object only functions when a man and woman touch each other. But why did He place the object in the woman for the man and in the man for the woman? If it would already be a mother within the object, He could make it potent or not. Therefore he made fantasy have the power to initiate the seed or not. If it is not carried away in the fantasy, it dies there; it goes through the fantasy and goes away with the fantasy. For if he himself were a complete mother capable of birth, having been divinely created this way, he would have to grow, but we are not going to concern ourselves with such a creation. For God wants man to be a man and a woman a woman and God wills that both are humans. [p. 257] He does not want them to be like a tree, in which man and woman are one thing. The sensitive creatures must have a sensitive counterpart. But what do we want to add to this? God did not want to have an insensitive (*ietelich*) human to propagate his kind, rather he wanted one without seed, that is why put the object in women, that one could accept or not; for God wants more than anything to maintain the purity of the light of nature. For in the pure there is the spirit and in the impure lies the dumb spirits. So, as we have said, we conclude with the beginning of the seed, what the mother of the seed is and how they are born. It should be clear now why God did not place the seeds within humans. It is like half a seed. And so if we were a theologian, as if we were philosophers, we want to give credence to our writing and not let things go without description. We recognize in Christ a philosophy of pure hearts and not tainted, one that wants a pure light and not a clouded version, a free and not an enslaved. We complain about theology, but we hope that philosophy will not exclude us.

Second Tract

First Chapter

As we intimated in the conclusion to the previous chapter, you should know that the nature of the seed, that is the essence of the seed and the nature itself are not together in the creation of humans, rather nature alone. [p. 258] Thereafter the essence of the seed, which is the speculation is born from the object as God created it: the nature for life in the body, in humans, and the object to the seed, the irrational things have this for themselves. Such an example suggests that an irrational being has the opposite of this system, which causes the seed to move . And so, the speculation in irrational animals determines their reality and their seed, and not their nature. Such a thing was placed in the object by God. But the rationality of spirit is only in humans and in no other being. We want to examine how humans are born, the material from which they come, the formation and growth of humans and this we will do in the following chapters.

Second Chapter

Here is the text about the birth of humans from humans. God put the material of the seed in human's will, which he took from the object, so when he has desire and his will is filled in the speculation, the *liquor vitae* (life liquid) heats up from the mind/thoughts (*speculation*) and becomes a seed of a new human. This is the case for men and for women. The life liquid becomes and remains a liquid – it does not become seed. But it will become a seed in the sense that remains separated from the body. For that reason all the nature, the guiding force, the spirits are in it. And in the same way the body is formed the liquid of life forms too. For the liquid of life is nothing but a hidden human. For the visible is the one who hides him. The liquid of life can best be compared to a shadow on the wall: it comes from humans and is formed to resemble the human. [p. 259] This is the same with the liquid, it is a microcosm, it is an internal shadow. But it has a substance, it is tangible, it has sensitivity and all the same things that form a body. It is the noblest in the whole body. However, it does not have a soul. It takes its life only from the human body and not from itself, but it is still like another human. Like one who looks in the mirror. So, it takes nature onto itself and it is like a human without sins.

Third Chapter

So therefore the will of humans causes desire in the object and then this liqueur turns into seed. It is the same as when the sun ignites a piece of wood or another fire burns wood from the excessive heat that moves it, but the flames are not moved. It is the same with the object, it ignited the other's microcosm, which burns and from it a seed emerges, like wood in a fire. This seed, which is born (as we have shown), remains motionless at his end destination, like the liquid of life it has its place in the body, the hand seed becomes hands, the foot seed the feet, the heart seed will become hearts, the brain seed brain and the same pattern continues for all body parts. And there is no particular place where the seed falls and remains; rather it remains stationary and hidden in its place, exactly where its liquid and material is. It separates itself from the liquid of life like froth from a soup or an impurity from wine. And the way that it lies there, it is known that if the seed does not pull towards itself, nothing would happen from the seed of this human. [p. 260] So, understand what we are trying to say to you. You would like to believe that this seed comes through the *vasa spermatica*, as foolish doctors profess, and then it comes through the *vasis spermaticis* when it is stimulated and it is this seed that they call sperm. This is the biggest lie that these doctors spread so that they can earn money and bolster their reputation.

Chapter 4

Now a little bit more about the sperm, that comes out of the vasis spermaticus. It is a *egestio* of the liquid of life, that cleanses itself... Therefore the sperma does not have no cleaning [sic], rather it is the case that when the impetus comes through the object and ignites the sperm that becomes seed. And so after the heating-up takes place, at that point the resulting *egestio* is thrown into the *vasa spermatica* and retains the seed. And the seed from which the human grows, that very same thing remains within the liquid and does not go into the sperm. We want to solve an argument for you, the one that blinded doctors want to propose and that which they use old philosopher's work to support. We are well aware that an image of a human emerges from the sperm, but this alone is not enough to make a human, but rather it makes an image. This very same thing does not rest in the place our opponents say that it should, for they are of the opinion that it will become a human, they are mistaken. They are misguided because the basis of their art and decisions lies in money, rather than in truth. Therefore take note of our solutions for the sake of argument. The sperm is therefore a kind of digestion from the *liquore vitae*, that it has power over the digestion; it can form itself in the way that the liquor deems in its nature. The main point is that the *liquor vitae* is part of the body; thus the sperm retains this kind of form within it and has power to form itself like its liquor. But you should pay attention to the reason that the sperm cannot be born in a way that deviates from the form that is within the liquor. [p. 261] This is the natural efficacy, there is no life, no sensitivity, no face, no ears, no taste, no smell. These pieces are all complete in the seed, without a doubt. That is why the sperm is a part of the *liquore vitae*, which is a microcosm. And so when it is heated it throws its waste (*egestio*) out, so the *etliches* remains in the liquor, that is the form, but nothing more.

Chapter 5

We want to turn out attention away from the above discussion and focus on the seed. As you heard in the previous discussion the sperm is just a waste product, not a seed. We will explain the seed in greater detail now. So give us your attention and understanding. We think that what we have said suffices. But understand seed as follows. We have concluded that the seed lies still after its heating in the place where it ends, for it is separated throughout the entire body and does not give birth to any human, the guiding principle (*Agstein)* alone brings it to the place of the birth, the place where the seed should be in its proper time. This process should be understood as follows: First notice that the man has half a seed and the woman has half; so the two make a complete seed. But the way they come together should be recognized as follows: In the womb (*matrix*) we find a pulling quality, which is like a magnet. This pulls the seed to it. Such a pulling strength within the mother comes and opens itself so that the seed can be born. At this time the work from the understanding that comes from the mind is used and the womb pulls the seed of the humours to it from the woman and the man, from hearts, from livers, from legs, from kidneys, from muscles, from blood, from skin and from the entire body. For whatever has a separate part in the body, it has its own seed. [p. 261] But the seed from each individual part, is only one seed when it all comes together. And therefore the womb pulls the right seed mixed with sperm from each person, and separates within it the seed and throws the sperm out and retains the seed. In this way the seed comes into the womb. And when the pulling power of the womb is not there, no birth will result. Because in this case sperm will enter but no seed. That is why we consider the doctors foolish, when they believe in the sperm as nature and want to make women fertile – this is all a lie. For when the pulling quality of the womb does not work naturally, the seed does not come to this place, and there will be no child. And if the sperm were seed, as they say, that would become a child without a mother, as they have often claimed, if this were so than she could grow children every day. But their lies are not well proven by them.

Chapter 6

As we have explained, the seed from humans goes into the womb. Now we want to tell you about the microcosm. When all the seed from all the parts come together in the womb, the nature of the womb organizes the seed of the head with the brain seed and the crystals[12] and the like in their place. The womb makes a map of how it should be arranged; like a carpenter places the wood in the correct order for a house. So at that point a seed would be situated in the mother, that means the microcosm; for there all things next to another belong to a human, except the life and the soul, which we will explain later. But to comment on the half seed, we want to note that the individual seed from one human is incomplete as far as becoming a human, on the grounds that God wanted to create humans from two, not from one and he wanted it to be placed in two, not one alone. [p. 263] We recognize philosophically such a reason. As a human comes from a human seed, no other being grows from humans; for without seed there would be no child. It is like a nut tree that grows from itself, that is why it is made from which it comes. Therefore all trees like grows from like, all the nut trees produce the same kind of nuts, nothing therein changes. Therefore it is the same with humans; if he were born from humans he would look like his father and mother. Therefore if there were only one kind of human, they would all look alike and would be of the same seed. But the mixing of seed from men and women yields variety, so that no human looks exactly like another. When seed is compared and uneven in scale, so that it never balances, that a human would want to be born that was the same as his father or his mother , as they then are.[13] Each human seed breaks the other's seed, that is, his likeness, and that is the main point, so that no human looks the same as another.

Chapter 7

When the womb has the seed within it, the natural process begins between the seed of the man and of the woman. Whoever is strongest and best is born, it is made like that one or naturally formed like that one or more parts. As an example: the seed from the brain of the man and thereafter from the woman: the two seed produce just one brain, but whichever of the two is strongest, that one's nature produces a similar looking child, that is much like them, but in no way completely the same. The main point is that a breaking occurs from the other seed which forever changes its shape. [p. 264] And further, that which is born, one from another, the more and bigger the break the more the variance in appearance. Therefore the seed from the nose builds the nose of the child, following the form of both parents and the mixing. Therefore one leg from the other, the other from another. That is why offspring have the same illnesses as the parent with the stronger seed. The more one seed conquers the other in its good birth, the more the child will be formed like that same person. But to talk about the birth of a boy or a girl, that is another discussion. Before that and the nature does things to the seed to make it into a child, as we have said, a process occurs and this we will explain. The womb pulls the seed into it, this is the first step in the process, in which seed of the other taints it, therefore whichever wins, not in the shape or outward appearance of the child, or its limbs, alone in the nature and that is a natural challenge that does not touch the seed. And this conquering occurs not from the strength of the seed, but rather from the strength of both natures, so the seed rushes into the mother. Therefore whichever seed reaches the womb first, this gives either a boy or a girl. If her seed reaches there before his it will be a girl, if his reaches there first it will be a boy.

Chapter 8

So back to our discussion of seed, which we have located in the free will of humans. Even though it is objectionable/disguisting to the *etlichen*, we are not going to worry about that. For the observation of one from the other does not lead to philosophy, and the same follows, the way the seed and the sperm are two different things; for where an *ietlicher* sperm should be a seed, then every day children would grow within pregnant women and elsewhere. [p. 265] Since the womb pulls toward itself only once and not more, recognize that nothing will become of the sperm. But we want to tell you more about the philosophy and we want to give you an overview. Since we have talked about the mother of seed, following the same logic one would think that as often as one has an object, that often seed would be sent forth; understand it in this way. As often as a man is overcome by an object in his will and he desires, that is how often he will have seed; it is the same with women. But the womb does not draw the seed toward it every time, it is therefore powerful. A pregnant women could, as we have shown, bear a seed anytime. But it does not come to fruition, where it should because the womb does not use its pulling power, the womb must remain quiet and complete and this occurs until she becomes too old, at that point nothing more comes out. For the pulling force dies out with age. Such tangential things are not far from the philosophy of birth. These things can be found in the physical presentation, for these things are more physiological than philosophical. Along those lines we discuss menstruation; from the following you should have a cursory understanding. Menstruation is not a flower or bloom of women, like a tree. You should know that menstruation is a waste product of mothers, it is the stuff from which they could procreate. The things that are unnecessary for conception and nourishment are thrown out every month. Just like the stomach that throws things out every day, and the liver and such things, the womb rids itself of the unnecessary things every month. And menstruation is nothing other than a waste product and not a flower. It is excrement of the womb like stool is the waste of the stomach and urine is renal waste and phlegm is brain or sinus waste and therefore other such things are like that as we find in the body, the *propriis morbis mulierum*.

Tract 3 [p. 266]

Chapter 1

Now it is necessary to talk more about seed and how it lies in the mother, as we have detailed above. First, it is very fragile and it can be damaged by chance when there is illness, just like a seed sewn in a field. In the same way and as soon as it is caught, there is much possibility for chance and many things can harm it, so that it is not caught. But these things we leave to physiology (*physic)*. Our text will conclude with a discussion of growth. So recognize with us that are many parts in the body that are from liver, from legs, from the middle part of the leg, from kidneys, as you know these things make a complete human. And as often as not, a seed comes from each of these parts. But you should also recognize in our text that there is liquor that the leg has within it, that the middle part of the leg has in it, the whole body, a liquor of the liver and of the kidneys. It follows that if a man's hand were cut off his child would not be without a hand, because the liquor of that hand was not taken, only the limb. But if the liquor were sick then we would see consequences. As with all illness that can be inherited through the liquor, these things are inherited by the child, except if the other seed overpowers it. Such things we leave to physiology and we return to our text in which we will discuss all the forms and deformities that children get and from whence they come.

Chapter 2

After we have spoken about what man and woman make, and we have said enough, we have only just begun to discuss the formation of children and we will say much more about that now. Know that the two seed do not receive the form, for it will be a human from it, that is nature. [p. 267] But if it is to be a man or a woman, notice that it must become one or the other. For one must conquer the other and the other must relent.

But as rarely happens, we notice the birth of hermaphrodites, people who are woman and man. But this is because there is an equal conquering of both seeds, so that neither of their features wins over the other; in this case there will be a hermaphrodite. But the difference is there; that is a hermaphrodite. Although he has both parts, but both attributes are not separated, so that he is neither completely a man, nor completely a woman. And therefore it is a defective birth, a mistake of nature, but not completely one or the other, but this seldom occurs.

But twins have such an origin. The matrix pulls the seed of woman and man towards it, but it pulls them incompletely towards itself, it pulls a piece of one part and thereafter t a piece and thereafter another; therefore as often as a piece comes off, that is how many children will be born. And this we need to understand as follows. When a matrix pulls, it draws towards her a seed like any other. But she can remain motionless, so that she only pulls the seed halfway and then it lies still, and remains so for an hour, until the pulling power is gone, then she pulls another toward her. Such double pulling makes two children. For the seed, if they are to be separated, they do not reconnect ever again. Here is an example that cannot be understood any different way, it is like when a man should pull a whole pile towards himself and because of weakness he only pulls a half, thereafter the other half part or the third part, one after another. Such a thing happens also in the mother, so if her pulling power is incomplete to take all in one attempt, she will pull once and then again and before her power is all gone, one more pull. [p. 268] Such an occurrence separates the nature of the pulling in a specific way. That is why twins or more children are born, the more the weaker. Although it does occur that twins are equal in size and this is because nature was as strong on both sides to foster their growth. For the size of the children and of the parents does not come from the material, as the rotten doctors say, rather the size or strength comes from the nature of the growing power in the characterists of the parents or from the mixing of their strengths and it follows that if the parents are healthy and strong or less strong the children will be too. We also do not want to contradict the case of the sperm[14] therefore: if it comes into the matrix by its own pulling power to give one half and leaves more behind, and still the strength of the mother does not let it go, it pulls so hard that all the others after it follow. For a singular seed is big enough for one child, or two or three or ten or twenty and has no consequence on the size of the child, only that the growing power is strong enough. If the seed that made two children large enough and complete in their growth; were to make one child, it would also not be any bigger.

Chapter 3

It is also important to recognize that children are born without complete bodies. For example: with two heads, with an extra hand, fingers and legs. The reason for this is the following: Nature is not complete in the *liquore vitae*. And where they are incomplete, the same limb may be thereafter become damaged and in many different ways. When a seed for the head breaks in two but all the other sees are singular than a child with two heads will be born. [p. 23] When the seed for fingers breaks in three, more or less, than the fingers will be ordered just like the seeds were broken. For the seeds do not always break in such a way to produce twins. Sometimes there are half twins and the like and infrequently there are deformed births. It also happens that the part that guides the seed forward is stopped either through sickness of the *liquour vitae* or through obstruction, or due to the weakness of the mother's attraction and in such a case this body part will not be given to the child. Such rare deformities of nature happen when the sperm is not immediately pulled in. When a seed breaks, or is incomplete then such monsters will be born: like children whose bodies have fused, or two children with one head and such things, as has been explained above. But in such a miscalculation of nature it must be recognized that twins are a normal growth from nature and from the seeds, so long as the growing power is complete and in the matrix, a child will be born. The other kinds of births arise from misuse. The misuse occurs at conception. When an order that contradicts nature occurs the seed will not be in the correct place to be received. So this seed works as hard as it can and from this deformity occurs. When it is ordered the way nature intended than a normal birth will occur, when this is not the case we see deformity.

Another kind of deformity occurs in the matrix, which we will explain. The womb pulls an unrecognizable seed toward it and from this unnatural monsters will be born, these are so offensive that we should not describe them. It also happens sometimes that an animal seed is eaten or drunk and when this seed makes its way into the body the womb pulls it towards itself. [p. 270] The result is a wondrous birth. Animals with sperm may also be eaten. When this comes in contact with a woman it is possible to grow something like it within her. Or another mixing that can occur: an animal can enter a mother and become mixed with the womb, this is possible. But such deformity is only in the sperm because that which comes from animals is only sperm. For in the animal one finds no *liquor vitae*. Animals have only seed in their sperm. We call this *semine animalium*. It is possible that such unnatural monsters can be born from the sperm of humans: if the power and the seed are retained in the body and the matrix pulls nothing from a man towards it, and it has only the seed from the woman than that can merge with the sperm and grow into an incomplete animal-like human.

For when one seed is not there and the other goes into the growing phase this results in a birth that is incomplete, or because of the sperm it may grow into something that is not recognized as human. These occurrences are common as the following will explain.

Chapter four

As we have concluded that seed lies in the mother, but now we want to leave that topic and explore how seed grows. God created a goal for the length of time necessary to grow a child: approximately 40 weeks: the same way for cows, doves and others – each has their own growing time. Now notice, however, that the time for a seed to become a child is thus: First comes the conception, as described above, after that, nature orders the seed the way they are supposed to be, the head seeds go to the head, the arm seeds go to the arms, the same for each part and its seed. [p. 271] When everything is in place the way it should be, the womb stops and does nothing more. Thereafter that material nature propels the child in its growth. When everything occurs as described above it grows into a body as it should. That which should make skin will become skin, that which belongs to legs will become legs, all limbs will become limbs and all internal organs will become the internal parts. And so the seed is no longer a seed, but rather blood and flesh as it should be in humans. After this phase the material nature dissipates and the grace of God completes the creation of humans and give life and all that belongs to life: seeing, hearing and the power to grow. But after life is given and nature is ordered by God, the body grows inside the mother's body until all the parts are complete and strong enough. After that, the spirit, soul, reason and understanding will be given to him and everything that goes with the soul. After that comes the time of birth – the soul and all that goes with it does not remain in the mother's body, it hurries out when the time is right. All this occurs in the above stated number of weeks and such births will be discussed further in *paramiris*,[15] in which the birth of the soul will get more attention. So you should understand how the received seed first forms and becomes blood and flesh, thereafter in the strength that he must endure the earth, it follows that he is given a spirit and soul. For when it is not strong enough it will succumb to the earth and it will not get a soul. When it is strong enough it gets a spirit and then the soul. [p. 272] First the spirit must be present in the body before a soul can enter, for the soul will not enter the body without a spirit. And the reason for this is that it is possible during growth that a child can become a monster, for which a soul is not appropriate. If the growth of the child is complete and not deformed, then the spirit will enter and it provides the way for the soul, then the soul enters. Thereafter the child grows in the unification of soul and body until it no longer gains nourishment from the mother and needs air, the birth follows this process.

Chapter five

In order to address a complete human and not only a child, it is important to recognize that when the child is given a soul all other things that complete a human are also given to him: vision, hearing, smell, touch, taste: the same with reason, knowledge, understanding and all the things that belong together. So as not to forget that we educate you, there are many humans who have a soul but do not have vision or hearing or something like that, like reason or knowledge. We want to educate you about these things so that you recognize that they are in every way human. The reason a person might lack something, like the ability to see, stems from a missing seed. God did not take this ability away, but nature caused this disability by failing to be pulled inward, or by being damaged in some way. But reason, on the other hand, cannot be taken away from a human, it is given to humans complete. If it is damaged it is because parts of the reason or senses are incomplete when nature put them there, they are unnatural. [p. 273] Although they are unnatural in the way that nature orders them, the spirit has difficulty affecting it. We need to recognize the smaller parts of the seed and how they function. We cannot see the seed and within it there is an invisible *liquor vitae*. This liquid may be damaged and a person with damaged *liquour vitae* may have bad seed. When the body of the seed is lazy, deaf and dumb the children will also be this way. But a clean fluid does not make a clean seed. From the clean seed, complete children will be born with all their limbs, though they may not have reason or the ability to see. But when these things are not in place, we will find no reason in the child. This is also the reason for lame and evil children. If the liquor is tainted, than the children will also be similarly corrupted. From whence the liquor comes, we recommend the study of physiology. With that we would like to conclude our discussion of the birth of humans.

Chapter six

We must say one last thing before we conclude this first book to teach you about the suggestion that many people talk about, namely how women themselves cause disability in their children by what they see. We want to give you an answer to this: The child is not changed by a woman's fantasy or what she sees, it will not be harmed by its mother. But she may encounter a shadow that scares her and causes her great astonishment or something so extreme that the blood that nourishes the child is drawn away from it or too much blood runs toward it and the nature of the mother in such a situation is altered. Through such a change the child becomes ill, but this illness appears to be greater than other human illnesses. For the child is delicate and the blood is scared blood that was tainted by the mother, this blood has the ability to change the child. The child may lack coloration, all colour will be drawn from it. We recommend that you study nature. Another thing that we want to tell you about is a child that grows in the form of a crab or a fish, an acorn or something like that. You should understand such births as follows: if a pregnant woman eats a crab and desires it. [p. 274] When the crab enters her body the nature of the mother's desire for it is so strong that the sperm comes with it adheres to the child as if to one of its own and the thing grows on the child and stays with it like a shadow on the wall. For all that the womb (*matrix*) wants to have enters into it, it does not go into the stomach like other foods, rather it enters the womb. Sometimes other foods go into the matrix and hurt what is in it rather than hurting the stomach. All this happens because the seed of the thing enters into the mother and becomes like a seed for growing life. This is an unnatural process. It can only be compared to a shadow. It leaves an impression on the child, but it is not really there. In the writing of astronomers they argue that stars and nature make humans, we will leave this viewpoint alone and recognize it as a good fable and a short joke that goes well with a good drink.[16] We conclude by calling their rationale a fable. We can think of many good jokes before the year comes to an end. There are many fools and they are hard to recognize.

NOTES

Introduction

1. I., J. and M. Millar (eds), 'Paracelsus', *Cambridge Dictionary of Scientists David* (Cambridge: Cambridge University Press, 2002), p. 281.
2. W. Pagel, *Paracelsus: An Introduction to Philosophical Medicine in the Era of the Renaissance* (1958; Basel: S. Karger, 1982), pp. 29–30.
3. The compilers, and Paracelsus scholars, K. Sudhoff and K. Goldammer both noted linguistic inconsistencies in the tract that did not seem to follow patterns found in other Paracelsus writing.
4. C. D. Gunnoe, 'Paracelsus's Biography Among his Detractors', in G. S. Williams and C. D. Gunnoe (eds), *Paracelsian Moments* (Kirksville, MO: Truman State University Press, 2002), pp. 3–17. Gunnoe outlines the sixteenth-century anti-paracelsian movement and how various detractors provided information for Erastus's vehement condemnation of Paracelsus's practice and character.
5. Pagel, *Paracelsus*, A. G. Debus, *The Chemical Philosophy: Paracelsian Science and Medicine in the Sixteenth and Seventeenth Centuries* (New York: Science History Publications, 1977). Debus's book focuses on Paracelsian scientists from the late sixteenth to the early eighteenth century. Other scholarship on Paracelsus's chemical philosophy includes B. T. Moran, 'Court Authority and Chemical Medicine: Mortiz of Hessen, Johannes Hartmann, and the Origin of Academic Chemiatria', *Bulletin of the History of Medicine*, 63 (1989), pp. 225–46. and 'Der alchemistisch-paracelsische Kreis um den Landgrafen Moritz von Hessen-Kassel (1572–1632)'. *Salzburger Beiträge zur Paracelsusforschung*, 25 (1987), pp. 119–45. Bruce Moran and William Newman also situated Paracelsus in the history of Science, but expanded our understanding of Paracelsus's contribution to the field of Alchemy, or early Chemistry see: B. Moran, *Distilling Knowledge: Alchemy, Chemistry, and the Scientific Revolution* (Cambridge: Harvard University Press, 2005) and W. R. Newman, *Atoms and Alchemy: Chymistry & the Experimental Origins of the Scientific Revolution* (Chicago, IL: University of Chicago Press, 2006).
6. Because of Paracelsus's wanderings, his manuscripts were scattered throughout Europe and finding them all still proves challenging. Paracelsus left writing behind wherever he went and seldom had the opportunity to revise his work. As a result, the majority of his works read as drafts in which he rethinks his stance and jots down ideas that he plans to explore later. Adding to the difficulty of dealing with Paracelsus's complicated writing, editors added their own organization and approximate dating to the remaining documents. First the works were collected and published by Johannus Huser in 1589.

The Huser edition of Paracelsus's works focuses on medical tracts. Similarly, the twentieth-century collection initiated by Karl Sudhoff also begins with medical works and is based on the Huser edition (*Paracelsus Sämtliche Werke*) Sudhoff intended *I*ntended to compile all of Paracelsus's medical writing in series one (fourteen volumes) and then to assemble all the theological writing in series two. He only finished the first volume of the second half of his series, but the work was later undertaken by Kurt Goldammer who followed the path set forth by Sudhoff and called his portion of the collection (the theological writing) series two. The vast array of theological tracts proved to be too much even for Goldammer to complete and the remaining tracts are currently being collated and published by Urs Leo Gantenbein.

7. K. Biegger, *'De Invocatione Beatae Mariae Virginis': Paracelsus und die Marienverehrung* (Stuttgart: Steiner, 1990) and U. Gause, *Paracelsus (1493–1541): Genese und Entfaltung seiner frühen Theologie* (Tübingen: J. C. B. Mohr, 1993). Biegger contextualizes Mary worship from the early medieval period up to and including Paracelsus' lifetime. She compares and contrasts the remaining copies of *De Invocatione Beatae Mariae Virginis.* Her conclusion brings into view the question of the overall image of women in Paracelsus's writing. She suggests that he should be credited for a generally positive view of women although he, at times, writes very negatively about them. Her analysis of the figure of Mary in Paracelsus's theological writings demonstrates that his understanding of her changes over time in his writings. Biegger suggests that Paracelsus may have read new Bible translations as well as writings of German Reformers, and these readings may have influenced his interpretation of Mary. Ute Gause, followed this path of inquiry further. Gause situated Paracelsus's writing on women in the context of women's subordinate role in society in Paracelsus's time. She postulated that all women in Paracelsus's time were subservient to men with the exception of nuns, who occupied a special status. Gause focused only on his early works (1520–5) and argued that women were primarily described in reference to men. She viewed the equality in reproduction outlined in his early works as representative of a larger assumption about women in general and noted that he probably granted women a more significant role in reproduction than the theologians or physicians of his time. Gause favoured Paracelsus's theological interests; thus she reads his medical tracts, such as *Volumen Parimirum* solely in terms of biblical interpretation.
8. For example, Katharina Biegger's 'Paracelsus über Maria-Weiblichkeit und Heiligkeit' *Nova Acta Paracelsia*, 14 (2000), pp. 3–18, p. 6ff explains how Paracelsus thinks through the meaning of Mary's body.
9. G. S. Williams, 'Die dritte Kreatur. Das Frauenbild in den Schriften von Paracelsus (1491–1543)' (1989) one of the earliest articles devoted to exploring Paracelsus's presentation of females from a feminist perspective. She situates his writing in a time of witch trials and overtly negative publications by Reformers about women's role in society demonstrating how Paracelsus went against the grain and portrayed a generally positive image of women. Gerhild Scholz Williams was my dissertation advisor and I greatly appreciate her support of my research and all the helpful suggestions she offered along the way. In addition to the authors already mentioned articles on Paracelsus's gender constructs by Porksön, Keller and Lundt have also added to our understanding of Paracelsus's writing on women. See: B. Lundt, 'Von Selbstgeburt und Spinnentod. Kontinuität und Kontrast der Genderbilder in erzählenden Texten über Paracelsus'. *Nova Acta Paracelsica*, 14 (2000), pp. 97–132, H. E. Keller, 'Zwo welt in einer haut: Paracelsische 'Augenlehr' am Beispiel der Frau', *Nova Acta Paracelsica*, 14 (2000), pp. 41–78 and G. Pörksen, 'Micro-

cosma. Der weiblich Leib. Zu einer Theorie der "Frauenheilkunde" bei Paracelsus', *Nova Acta Paracelsica*, 14 (2000), pp. 79–96. Pörkson's article is one of few that brings attention to Paracelsus's sexed-based medical practice.

10. While I was working on this project Gunhild Pörksen also recognized the value of studying how Paracelsus interpreted the female body when she focused on the female body as microcosm. Pörksen, 'Microcosma'.
11. A. Weeks, *Paracelsus: Speculative Theory and the Crisis of the Reformation* (Albany, NY: State University of New York Press, 1997). Weeks's book focuses on the similarity between Luther and Paracelsus. Because of this emphasis, attention to Paracelsus's medical interests is limited. The book provides an excellent introduction to Paracelsus's theological interests in the context of the Reformation, but the impact of the Reformation on his approach to medicine does not get as much attention as it deserves. Weeks attempts to introduce readers to Paracelsus by describing him as a radical Reformer, who throughout his writing drew on the religious interests of his time. I agree with his interpretation, but the broad scope of his analysis downplayed Paracelsus's importance in the field of medicine, the profession with which Paracelsus strongly identified throughout his life. Charles Webster's approach to Paracelsus, like Weeks, took into account the interplay of religious and medical thought in Paracelsus's writing, but also suggested connections between his writing and the limited biographical information available to us. C. Webster, *Paracelsus: Medicine, Magic and Mission at the End of Time* (New Haven, CT: Yale University Press, 2008).
12. W. Pagel, 'Paracelsus and the Neoplatonic and Gnostic Tradition', *Ambix*, 8 (1960), pp. 125–66, p. 135. Pagel attempted to situate Paracelsus's understanding of the body in neo-platonism and his article makes the similarities clear. Pagel was undoubtedly correct in pointing out how sexual duality permeated Paracelsus's writing, but this study takes a different approach to Paracelsus's inspiration, namely his quest to carve out his place in the emerging medical community using biblical interpretations as the basis of his practice and his model of understanding the body. I do not try to situate Paracelsus's contemplation of the body in light of classical texts, because he repeatedly eschewed such study. Instead, I follow Paracelsus's engagement with the topic of conception and gestation arguing that his quest to define his theory of conception ultimately led him to a theory of embodiment and that influenced future medical practice and the way that we think about the modern body.
13. This topic has not been considered in larger works on Paracelsus, but has recently gained attention a few articles. I found reference to Paracelsus's formulation of a sex binary in Hildegard Elisabeth Keller's article, 'Seeing Macrocosma: Paracelsus's Gendered Epistemology', in G. S. Williams and C. D. Gunnoe, *Paracelsain Moments* (Kirksville, MO: Truman State University Press: 2002), pp. 93–116. Keller focuses on Paracelsus's tract De Caducus Matricis.
14. T. Laqueur, *Making Sex: Body and Gender from the Greeks to Freud* (Cambridge, MA: Harvard University Press, 1990), p. 149.
15. For example, Katherine Park explained the sixteenth-century French medical writing about the clitoris refuted Laqueur's claim by showing that physicians did have terminology for female anatomical parts (a point Laqueur denied) and that the discussions about the clitoris implied a two-sex model of the body in: 'The Clitoris Rediscovered', in D. Hillman and C. Mazzio (eds), *The Body in Parts* (New York: Routledge, 1997), pp. 171–94. M. Stolberg, 'Critiques and Conventions: A Woman Down to Her Bones', *Isis* 94:2 (2003), pp. 274–99 argues that the female skeleton and explicit sexual dimorphism

appeared in the sixteenth century, namely in 1583 in Felix Platter's *De corpus humani structura et usu libri III*. Stolberg notes that Platter's work was remarkable in that it labelled the anatomical representation of the female body to show how it differed from the male body. Stolberg has demonstrated that some of the core aspects of Laqueur's work arguing for the development of a two-sexed model of the body actually occurred in the sixteenth century, specifically in 1583. Several other studies of anatomy in the early modern period were undertaken shortly after the publication of Laqueur's book: A. Carlino, *Books of the Body: Anatomical Ritual and Renaisssance Learning*, trans. J. Tedeschi and A. Tedeschi (Chicago, IL: University of Chicago Press, 1999), J. Sawday, *The Body Emblazoned: Dissection and the Human Body in Renaissance Culture* (London: Routledge, 1995) and D. Hillman and C. Mazziod eds. *The Body in Parts: Fantasies of Corporeality in Early Modern Europe* (New York: Routledge, 1997). Studies on who had authority to practice anatomy have also gained attention of late. For example, the dissection by and of nuns in fourteenth and fifteenth century Italy revealed expectations that within the female body one would find visual signs of the divine. In other words, the female body was unique in that it had the ability to show a woman's virtue through signs on her internal organs. K. Park illuminated this aspect of dissection in her study of late medieval Italy. Her research focused on trials for sainthood. Other research on how the body was understood in medieval Europe also points to a distinct and separate understanding of the workings of the female body. See: *Secrets of Women: Gender, Generation, and the Origins of Human Dissection* (New York: Zone, 2006) and 'Relics of a Fertile Heart: The "Autopsy" of Clare of Montefalco'. *The Material Culture of Sex, Procreation, and Marriage in Premodern Europe*, in A. McClanan and K. Rosoff Encarnación (eds), (New York: Palgrave, 2002), pp. 115–33

16. M. Fissell, *Vernacular Bodies: The Politics of Reproduction in Early Modern England* (Oxford: Oxford University Press, 2004). She notes how women were no longer encouraged to pray to Mary for assistance in childbirth and how women's knowledge about reproduction was challenged, and later downplayed, as a result of the Reformation and the civil war in England.
17. A. Cooper explains Paracelsus's important role in pharmaceuticals and the development of a local (Indigenous) categorization of botanicals in *Inventing the Indigenous: Local Knowledge and Natural History in Early Modern Europe* (Cambridge: Cambridge University Press, 2007), pp. 22–32.
18. K. Crowther-Heyck, '"Be Fruitful and Multiply": Genesis and Generation in Reformation Germany', *Renaissance Quarterly*, 55 (2002), pp. 905–34
19. A. Fausto-Sterling, *Sexing the Body: Gender Politics and the Construction of Sexuality* (New York: Basic Books, 2000).
20. This may be confusing for readers who are familiar with medieval and early modern cosmology. At this point I am not referring to the body as a microcosm of the greater would, but rather the modern scientific idea that the body's sex can be understood on a minute scale based on hormones, chromosomes or adrenal function.

1 Rethinking Sex in Early Modern Germany: Negotiating Medical Authority

1. M. Green's *Making Medicine Masculine* (Oxford: Oxford University Press, 2008) writes about the theory and practice of gynaecology in medieval Europe by explaining how writing about the female body and the process of conception of gestation was a mode for men to gain authority in the field. Green explained the transition from women's authority in the medieval period (as exemplified in the widely-circulated, woman-authored *Trotula*) to the way in which the field became dominated by men. She argues that literacy played a part in elevating the status of male physicians and that the revival of Hippocrates gynecological texts in the sixteenth century underscored the legitimacy of male practitioners in the field of gynecology. Her study focuses on the practice of gynecology.
2. P. O. Kristeller's book, *Renaissance Thought and Its Sources* (New York: Columbia University Press, 1979) provides a more detailed analysis of how medieval engagement with classical texts influenced Early Modern thought.
3. Understanding the body as a discourse emerged from the work of M. Foucault, *Discipline and Punish: the Birth of the Prison*, trans. A. Sheridan (New York: Vintage Books, 1995), *The History of Sexuality* trans. R. Hurley, 3 vols (New York: Vintage Books, 1990).
4. Leah DeVun explains that Petrus Bonus of Ferrara explicitly connected the idea of hermaphorditism to the alchemical agent of transmutation in 'The Jesus Hermaphrodite: Science and Sex Difference in Premodern Europe' *Journal of the History of Ideas*, 69:2 (2008), pp. 200–1.
5. William R. Newman describes this medieval alchemical philosophy in: *Atoms and Alchem*, p. 23–44.
6. C. W. Bynum, *Fragmentation and Redemption* (New York: Zone Books, 1991), p. 92 Bynum's essay, 'The Body of Christ in the Later Middle Ages' is particularly salient for this study. The analysis of visual representation of Christ provides a pathway for questioning the way the body and sexuality were understood in the later middle ages. In particular, Bynum writes, '... there is both iconographic and textual evidence for the argument that late medieval people sometimes saw the body of Christ as female'. Furthermore, she draws connections between medieval medical understanding of the purging of blood and Christ's blood on the cross. Both, she argues, derive from a medieval understanding of the female body. Bynum suggests that women authors played an important role in disseminating such ideas. She cites, for example, Mechthild von Magdeburg (1207/10 *c.* 1282 or 1297). See: C. W. Bynum, *Jesus as Mother* (Berkeley, CA: University of California Press, 1982), pp. 228–47 and *Fragmentation and Redemption* p. 101
7. Laqueur, *Making Sex*.
8. I use the phrase 'Germany' to describe regions of the Holy Roman Empire where German was the primary language.
9. We know that he was born in Einsiedeln, located near Zürich, Switzerland around 1493 and was given the name Theophrastus Bombastus von Hohenheim, but began signing his works Paracelsus around 1529. This transition to the signature Paracelsus is an important point to remember because other writers of the day refer to him as Theophrastus, as Hohenheim and also sometimes as Paracelsus. There are various explanations for the origin of the name. Walter Pagel suggested that it could either be a translation of Hohenheim, the famiy estate in Swabia, or it was intended to signify the surpassing of Celsus, the Roman writer on medicine, or possibly Celsus, the enemy of the Church. (p. 5). Andrew Weeks noted that Paracelsus also titles many of his tracts 'Para—'. (p.

3). Paracelsus undoubtedly learned a bit of the trade from his father during his youth in Villach, located in present-day Austria, where his father worked as a physician. His own formal education in the field began in Vienna and continued in 1512 in Italy, where he matriculated at Ferrara and subsequently passed examinations in the two arts of healing (*Leib-und Wundarznei*). We do not have documentation to confirm that he graduated from the medical programme.

10. The often cited story of his time in Basel is as follows: as a professor, Paracelsus was equally as unconventional as he was as a student. He refused to teach after Galen and Hippocrates, and instead wanted to share his experience as a wandering physician with students (in Latin and German) for two hours a day. At this time he wrote about stone-forming illnesses, but he did not stay in Basel and work on this project for long. As if his unorthodox choice of lectures were not enough to instill hostility, he publicly burned the 'Avicenna', the canon of academic medicine. For this, the faculty removed his privilege to use the lecture halls and refused him the right to sponsor students for the doctorate. He retaliated by lecturing only in German, which drew larger than usual crowds including barber-surgeons. When his sponsor, Froben, became seriously ill Paracelsus's medical practices were put into question.
11. Following many years of wandering, in 1540 he received a call to Salzburg from Archbischop His Ernst von Wittelsbach. On 24 September 1541 he died and was buried there in Saint Sebastian, a cemetery for the poor. Portions of this biography were extracted from Udo Benzenhöfer's *Paracelsus* (Reinbeck bei Hamburg: Rowohlt, 1997), Helga Naber's *Probleme einer Paraclesus-Biographie* (1998), and Walter Pagel's *Introduction to Paracelsus*.
12. See: C. Merchant, *The Death of Nature* (New York: Harper and Row, 1993). Katherine Park argued that Merchant overstated the use of Nature as Woman and argued that in the medieval period the image of nature was primarily androgynous, but that it emerged as female in the early modern period. Park draws attention to Robert Fludd's Utriusque cosmic maioris scilicet et minoris metaphysica, physica atque technical historica (*c.* 1617) to describe the Early Modern concept of nature as a microcosm. She overlooks the influence of Paracelsus on Fludd (a known follower of Paracelsus) in 'Nature in Person: Medieval and Renaissance Allegories and Emblems', in L. Daston and F. Vidal (eds), *The Moral Authority of Nature* (Chicago, IL: University of Chicago Press, 2004), pp. 50–73 William R. Newman disagrees with Merchant on other grounds: He argues that she misunderstood the history of alchemy and that reverence for nature (symbolized as a female body) continued well into the modern era. I agree with his conclusion, but question his use of the tract *De Natura Rerum* to explain Paracelsus's viewpoint because the text may not have been written by Paracelsus. Karl Sudhoff pointed out many inconsistancies with this tract. Newman's article, ‚Alchemy, Domination and Gender' can be found in N. Koertge (ed.), *A House Built on Sand: Exposing Postmodernist Myths about Science* (New York: Oxford University Press, 1998), pp. 216–26.
13. Park, 'Nature in Person'.
14. Although her focus is on the seventeenth century, L. Roper, *Witch Craze: Terror and Fantasy in Baroque Germany* (New Haven, CT: Yale University Press, 2004) provides excellent background on early discussions of witchcraft in sixteenth-century Germany. The book includes reproductions of woodcuts dating from the sixteenth century that show the suspicion of women congregating in areas exclusive to them, in particular bathhouses; but Roper explains throughout her study that the lying-in room was generally held to be the most suspicious place of all. There are numerous studies on witchcraft

available now. In addition to those already mentioned, the following are also relevant to understanding the phenomena in Germany: W. Behringer (ed.), *Hexen und Hexenprozesse* (München: Deutscher Taschenbuch Verlag, 1988), B. Ankarloo and S. Clark (eds), *Witchcraft and Magic in Europe: The Period of the Witch Trials* (Philadelphia, PA: University of Pennsylvania Press, 2002), especially part 1, pp. 1–49.

15. Gerhild Scholz Williams has argued that Paracelsus's approach to witchcraft did not implicate women to the same extent as other tracts on witchcraft from the period. Unlike many of his contemporaries who believed that coitus with the devil led one to witchcraft, Paracelsus believed that one was predestined to witchcraft. The ascendants (Paracelsus defines these as planetary spirits) entered the body at the moment of birth. See: G. S. Williams, *Defining Dominion* (Ann Arbor, MI: University of Michigan Press, 1995), pp 45–63.
16. U. Rublack, 'Pregnancy, Childbirth and the Female Body in Early Modern Germany', *Past and Present*, 150 (February 1996), pp. 84–110
17. See for example, S. C. Karant-Nunn and M. Wiesner-Hanks, *Luther on Women* (Cambridge: Cambridge University Press, 2003) and E. Labouvie, *Andere Umstände* (Köln: Böhlau, 2000).
18. J. Spinks, 'Wondrous Monsters: Representing Conjoined Twins in Early Sixteenth-Century German Broadsides', *Parergon*, 22:2 (2005), pp. 78–112.
19. A great deal of scholarship has explored the way early modern viewers of these text interpreted the abnormal births. Spinks suggests that the early sixteenth-century showed more compassion toward the deformed child than the texts of the second half of the century. Spinks, 'Wonderous Monsters'. Previous scholarship focused on whether these broadsides understood the monsters as natural or supernatural occurrences, as emblems or as symbols. See L. Daston and K. Park, *Wonders and the Order of Nature 1150–1750* (New York: Zone Books, 1998), M.-H. Huet, *Monstrous Imagination* (Cambridge, MA: Harvard University Press, 1993) and A. Bates, *Emblematic Monsters* (Amsterdam: Rodopi, 2005). See for example, Labouvie, *Andere Umstände*, M. E. Wiesner, 'The Midwives of South Germany and the Public/Private Dichotomy', *The Art of Midwifery: Early Modern Europe* (London: Routledge, 1993), pp. 77–94. Lynne Tatlock's translation of the work of Justine Siegemund introduces English readers to a primary text by a female midwife of the seventeenth-century: J. Siegemund, *The Court Midwife* (Chicago, IL: University of Chicago Press, 2005).
20. Laqueur, *Making Sex* pp. 38–43, M. Wiesner, *Women and Gender in Early Modern Europe* (Cambridge: Cambridge University Press, 2000), pp. 30–3, J.Cadden, *Meaning of Sex Difference in the Middle Ages (*Cambridge: Cambridge University Press, 1993), Green, *Making Women's Medicine Masculine.*
21. This debate between an Aristotelian model *vs* a Galenic model has a long history. For an overview of the topic in the medieval period see: Joan Cadden. *Meanings of Sex-Difference in the Middle Ages: Medicine, Science and Culture* (Cambridge: Cambridge University Press, 1993) and Merchant, *The Death of Nature, Women* also shows the variance of discussions about human reproduction in the early modern period revealing that scholastic debate on the issue contained elements of both Aristotelian and Galenic philosophy. In other words, there was no consensus regarding human conception. For more on this topic see especially pp. 156–8. Merchant's book provides an overview of how discussions of nature in the early modern period and beyond tended to incorporate concepts of femininity and the female body. Eva Labovie's work also discusses the German context. See: Labouvie, *Andere Umstände*. For more information on how the

unborn child was understood in this period see: B. Duden, 'The Fetus on the "Farther Shore": Toward a History of the Unborn', in L. M. Morgan and M. W. Michaels, *Fetal Subjects, Feminist Positions* (Philadelphia, PA: University of Pennsylvania Press, 1999), pp. 13–25.

22. Ryff (same as above) page 3v. translation mine. Other authors refer to Paracelsus as Theophrastus at this time; however there is a possibility that Ryff refers to the Greek philosopher Theophrastus who was known for herbal remedies. The grouping of this passage with references to the kind of doctors who interfere with birth, leads me to assume that he was condemning Parcelsus's viewpoints.
23. For more about Paracelsus's name change see: Webster, *Paracelsus*, p. 15
24. M. J. O'Dowd, *The History of Medications for Women* (New York: Parthenon, 2001), pp. 141–52.
25. E. Rößlin: Der schwangerenn Frawen und Hebammen Rosengarten, Augspurg, 1530 [VD16 R 2856] access through Bayrische Staats Bibliothek: http://mdz10.bib-bvb.de/~db/0002/bsb00027287/images/index.html [accessed 15 December 2008].
26. K. Crowther, '"Be Fruitful and Multiply": Genesis and Generation in Reformation Germany', *Renaissance Quarterly*, 55 (2002), pp. 905–34, p. 905.
27. I am grateful to Kathleen Crowther, who has researched Ryff's writing and the visual representations present in his work. I shall draw on her research when exploring how Paracelsus understood the human body as analogous to the biblical figures Adam and Eve in chapter four.
28. Walter Herman Ryff, *Frauen Rosengarten Von Gebrechen der Mutter und Kinder* Frankfurt 1547. Access online through Bayrische Staats Bibliothek: http://daten.digitale-sammlungen.de/~db/0002/bsb00029622/images/ [accessed 15 December 2008]. This reference is on page 2R of the original text. This section of the book has many marginal glosses next to each reason for infertility.
29. Crowther, '"Be Fruitful and Multiply"'. She suggests that Paracelsus's writing was idiosyncratic of the period however this study should make clear that he ascribes to some of the same metaphors and symbols of the Fall that she recognized in the work of other authors of the period.
30. Wiesner, *Women and Gender in Early Modern Europe*, p.21
31. See: Karant-Nunn and Wiesner-Hanks, *Luther on Women*, Peter Brown, *The Body and Society: Men, Women, and Sexual Renunciation in Early Christianity* (New York: Columbia University Press, 1988), Ulrich Asendorf, Lectura in Biblia: Luthers Genesisvorlesung (1535–1545) (Göttingen: Vandenhoeck & Ruprecht, 1998).
32. In Paris in 1271 the faculty of medicine wrote a document prohibiting Jewish doctors from practising medicine or surgery on anyone of Christian faith. See: J. Shatzmiller, *Jews, Medicine and Medieval Society* (Berkeley, CA: University of California Press, 1994), p. 16.
33. In Basel, where we know that Paracelsus briefly practiced, Jews were prohibited from earning university degrees beginning in 1434. See: V. Bullough, *The Development of Medicine as a Profession* (New York: Hafner, 1966), p. 107.
34. Udo Benzenhöfer and Karin Finsterbusch examine anti-Semitism in the works of Paracelsus in their article, 'AntiJudaismus in den Medizinish-Naturwissenschaftlichen und Philosophischen Schriften des Paracelsus', in H. Schott and I. Ziguer (eds), *Paracelsus und Seine Internationale Rezeption in der Frühen Neuzeit* (London: Brill, 1998), pp. 96–109. They approach the issue by considering how anti-Semitism was different in the early modern period, than in the modern era when Paracelsus's writing was beloved by

National Socialists such as Sepp Gottlieb. Their list of differences is both troubling and curious as it seems to heighten the process of bigotry while deconstructing it. Despite this methodological flaw, the authors tease out the instances in which Paracelsus writes against the Jews. Many of their references were from Paracelsus's late work, *Astronomia Magna*. In this sense, they fall into the trap that many previous scholars have: using the late works to try to understand earlier references. In this late work, Paracelsus considers how Jews are damned because of their beliefs (or lack thereof). The authors differentiate the theological concerns from Paracelsus's stance against Jewish doctors. In some cases, they argue, he conflates their religion with their medical skill, in other cases he writes against their ability to read Hebrew (p. 105). Though the authors do not interpret this segment as part of Paracelsus's general approach to medicine, I think it is clear that he opposes their ability to use sources from other languages in the same way he opposes those who relied on Latin medical texts.

35. This was the case, particularly of Jews in Spain because they gained access to (and could read) Arabic text as early as the thirteenth century. See: N. Sirasi, *Medieval and Early Renaissance Medicine: An Introduction to Knowledge and Practice*, (Chicago, IL: University of Chicago Press, 1990): pp. 29–31. It is important to outline the practice of Jewish physicians because Paracelsus cautions against their practices in several of his tracts. Although these passages have been interpreted as evidence of Paracelsus's anti-Semitism, it should be emphasized that he does not criticize their religious beliefs in defining how a doctor should practice, even though he emphasizes the spiritual quality of doctoring. Another part of the history of Paracelsus and anti-Semitic rhetoric of the early modern period, actually stems more from Paracelsus's followers than from Paracelsus himself. Scholars have shown that paracelsian physicians of the seventeenth century sought legal measures against Jewish physicians and claimed that their practices were far inferior. See: J. Levi, 'Paracelsians versus Antiparacelsians: German Doctors against Jewish Physicians in Frankfurt am Main 1618–1645', in H. Schott and I. Ziguer (eds), *Paracelsus und Seine Internationale Rezeption in der Frühen Neuzeit* (Leiden: Brill, 1998), pp. 110–30.
36. M. Walton, 'Anthonius Margaritha – Honest Reporter?', *Sixteenth Century Journal*, 36:1 (2005), pp. 129–41.
37. Of course, Luther warned against Jews in general, not just Jewish physicians. I am arguing here that Paracelsus borrowed the rhetorical strategy of Luther. The ties between Luther and Paracelsus have been examined in another context. H.C. Erik Midelfort, argued that Luther's approach to mental illness echoed Paracelsus's understanding of it and that Luther also understood the tripartite nature of the body as Paracelsus explained it in: *A History of Madness in Sixteenth Century Germany* (Stanford, CA: Stanford University Press, 1999), pp. 80–137. Midelfort's chapter illuminates the tension between medicine and theology when he describes how Luther was critical of physicians who placed too much emphasis on natural causes and warned that such an approach failed to recognize the power of Satan and God.
38. Andrew Weeks argued that we should understand Paracelsus as a reformer of medicine in much the same way Luther reformed the Church in his book, *Paracelsus: Speculative Theory*.
39. The way in which Paracelsus engaged in explaining his understanding of conception as a way to reinforce his medical authority is similar to the writing of male-midwives in late sixteenth-century France. In *Childbrith and the Display of Authority in Early Modern France* (Burlington, VA: Ashgate, 2005), Lianne McTavish shows how writing on obstet-

rics differed based on how the author identified, either as surgeons, physician or midwife. The obstetrical texts, she argues, served as a way to self-identify.

40. V. Nutton, 'Medicine at the German Universities, 1348–1500', in A. Cunningham and R. French (eds), *Medicine after the Black Death* (Cambridge: Cambridge University Press, 1995), p. 173.
41. Paracelsus's father studied medicine at the University of Tübingin. That university was founded in 1477. See: V. Nutton, 'Medicine at the German Universities, 1348–1500', *Würzburger medizinhistorische Mitteilungen*, 16 (1997), pp. 173–90.
42. Nutton notes that the statues of Leipzing and Wittenberg (of 1502 and 1508 respectively) requested, 'more modern' commentators such as Giovanni Arcolano (*c.* 1390–1458) and Jacques Despars (*c.* 1380–1458). However, there was a surge in interest in Latin editions of Avicenna's Cannon between 1470 and 1540 and editions of Aristotles works continued to be printed during that time, but reached its peak after about 1535. See: Midelfort, *A History of Madness*, p. 148.
43. K. Park, 'The Life of the Corpse: Division and Dissection in Late Medieval Europe', *Journal of the History of Medicine and Allied Sciences*, 50 (1995), pp. 111–32.
44. Nutton, 'Wittenberg Anatomy', p. 13.
45. P. F. Grendler, 'The Universities of the Renaissance and the Reformation'. *Renaissance Quarterly*, 57:1 (Spring 2004), pp. 1–42.
46. It is important to note that various editions of the works of Galen were accessible to German speakers after they were published in Basel in the sixteenth century. See: A. Wear, 'Medicine in Early Modern Europe 1500–1700', in L. Conrad (ed.) *The Western Medical Tradition 800B.C. to 1800* (Cambridge: Cambridge University Press, 1995), p. 253.
47. E. Bonjour, *Die Universität Basel von den Anfängen bis zur Gegenwart* 1460–1960. (Basel: Helbing und Lichtenhahn, 1971). Bonjour understands the beginning of medical study in Basel with Paracelsus and jumps to Vesalius with no mention of any possible lectures in between. Katherine Park explores the images in Vesalius's book as representitive of the male scientific discourse associated with conquering the female body in 'Dissecting the Female Body: From Women's Secrets to the Secrets of Nature', in J. Donawerth and A. Seeff (eds), *Crossing Boundaries: Attending to Early Modern Women* (Newark, DE: University of Delaware Press, 2000), pp. 29–47.

2 Predestined Conception: Seeds of Procreation and the Workings of the Womb

1. See the Appendix for a full translation of this tract.
2. Paracelsus, *Paracelsus Sämtliche Werke. I. Abteilung. Medizinische, naturwissenschaftliche und philosophische Schriften*, ed. K. Sudhoff, vols 6–9 (München/Berlin: O.W. Barth, 1922–5) vols. 1–5 and 10–14 (Munich: R. Oldenbourg, 1928–33) (hereafter *Paracelsus Sämtliche Werke I*), pp. 242–83. The dating of this work, and other early works, remains uncertain. However, it is clear from the rudimentary expressions of understanding the process of conception that Paracelsus is only beginning to engage with the topic, as the remainder of the chapter shall make clear. For an overview of the problems in dating the early works see Weeks, *Paracelsus: Speculative Theory*, pp. 33–47 and pp. 88–9
3. His major complaint was that such philosophers (Aristotle in particular) reasoned only intellectually neglecting observation, '*das sie aus iren sinnen geschriben haben und nichts aus der erfarenheit und auglichem anzeigen. Alein die erfarenheit und wissen sol schreiben*

und nicht das gedünken und vermeinen erfaren sein'. '...because they wrote only from their reasoning and not from visual experience. Only experience and knowledge should be written and not thoughts and opinions' *Paracelsus Sämtliche Werke I*, p. 243

4. He argues that Aristotle and his followers did not recognize the *light of nature*. Paracelsus contrasts his own reasoning with what he characterizes as ancient pondering, and he begins his inquiry by understanding the patterns of nature and using observation as the basis for his philosophy. Nature should be understood in context of the early modern period. Kurt Goldammer describes the varying definition, '*Das, was wir heute "Natur" nennen, wird im Mittelalter und in der frühen Neuzeit als Schöpfung, als Universum, allenfalls als Kosmos (vor allem in dem polaren Paar Makrokosmos und Mikrokosmos) bezeichnet. Damit ist eine Beziehung zur Religion, zum Weltbild gegeben, während die heutige Begriffshypostasierung einer alles beherrschenden Größe "Natur" fehlt*'. 'What we today call nature was described as creation in the middle ages and in the early modern period. It was a universal, a Cosmos (described primarily as a polar pair: macrocosm and microcosm). At that time, nature related to religion and a concept of the world, whereas today's concept of nature as its own force was missing from the medieval and early modern thinking'. Essentially there was no concept of an idea of nature that was not controlled by/influenced by God. See: K. Goldammer, *Der Göttliche Magier und die Magierin Natur* (Stuttgart: Franz Steiner, 1991), p. 11. It is for this reason, argues Goldammer, that Paracelsus was interested in the 'Magicians' of the New Testament, such as Matt. 2. Gause, in *Paracelsus*, argues that Paracelsus's concept of the Light of Nature changes in his later works.
5. He writes, 'aber die ursachen darumb wir den zweifüßigen menschen zu einem exempel <nemen>, ist nit die darumb das wir ein solcher sind oder darumb das unser geblüt darin erfunden ist oder das uns bekant ist mer dan andern oder darumb das dise secten den größesten wissentlichsten teil der welt besizt; nein, deren keins nit. Aber das ist die ursachen, das wir uns gedünken etwas und nit unzimblich, dieweil Christus unserer secten gewesen ist und unser form angenommen, der da die wal gehabt het under allen oder ein besondere zunemen, und sich in die unsere ergeben on natürlichen samen, achten wir das die andern secten nit volkomen in der form sind, sondern etwas bresthaftiges oder spöttisch in irer form, welchs nit in der form der zweifüßigen nit gebrist. [...]' (*Paracelsus Sämtliche Werke I*, p. 248).
6. Der ander ist, das etlich geschöpf sind als die entpfintlichen, die da in inen keinen samen haben entspringend aus irer natur, sonder ganz on allen samen leben und sind; das sind die menschen und dis vich. Kein man hat in ihm den samen, kein frau, der inen aus irer natur entsprünge oder geboren würde als in den unentpfintlichen, desgleichen des ander vich. (*Paracelsus Sämtliche Werke I*, p. 252)
7. In *Paracelsus Sämtliche Werke I*, p. 249 he mentions the eternal qualities and writes about the reasons God created humans this way, 'darumb hat er im den samen genomen aus der natur und im nit eingebildet von wegen des ewigen so in im ist. Denn er wil einen lautern menschen haben und nit ein verenderten, als der same tut so er in der natur ist' (*Paracelsus Sämtliche Werke I* p. 254).
8. He defines humans as 'von einem got, der die sêl gibt in allen' (*Paracelsus Sämtliche Werke I*, p. 247).
9. 'kein man hat in im den samen, kein frau, der inen aus irer natur entsprünge oder geboren würde als in den unentpfintlichen...' (*Paracelsus Sämtliche Werke I*, p. 252).
10. Warumb reden dan die blinden arzte von vil capiteln de spermate, also ob sperma der same sei, und ist kein sam. Wollen durch ire kunst, recepten und suppenwerk die frauen

geberhaft machen, den sperma generiren, das in und iren künsten nicht müglich ist. Wan es ist night materialsch, alein speculz (*Paracelsus Sämtliche Werke I*, p. 255).

11. 'Buch von der Gebärung', (*Paracelsus Sämtliche Werke I*, p. 254).
12. Ibid., p. 255.
13. Ibid.
14. Got hat dem menschen gesetz die materiam des samens in seinem willen den er im nimpt aus dem object. Also so er nun das verhengt hat und sein willen erfüllet in der speculation, als dan entzünt sich der liquor vitae von der speculation und wird zu einem samen eines andern menschen. Also ist dis im mannen und also auch in frauen. (*Paracelsus Sämtliche Werke I*, p. 258)
15. *Paracelsus Sämtliche Werke I*, p. 258.
16. Ibid., p. 259.
17. Ibid., p. 259.
18. As we shall see, in other tracts, like *Volumen Paramirum,* he is more concerned with how a human gains a soul.
19. Also verstanden uns, das ir möchten gedenken in euch, das diser geborne same keme in die vasa spermatica, von welchen die narrechten arzet redden, und als den vasis spermaticis selbs durch eigen anreizung heraus kem und also der same, den sie sperma nennen, heraus kom und also verschüt. Das ist der größisten lügen eine so die arzet gepflogen, ir gut zu gewinnen, und ire hofart zu erzeigen und die leut zuessen. (*Paracelsus Sämtliche Werke I*, pp. 259–60)
20. *Paracelsus Sämtliche Werke I*, p. 260.
21. He expresses his dislike for the profitable practice, 'dan ire gedanken und kunst ist mer auf gelt dan uaf die warheit gericht'.
22. *Paracelsus Sämtliche Werke I*, pp. 260–1.
23. Ibid., p. 261
24. Ibid., p. 262.
25. Ibid., p. 261.
26. It should be noted that Paracelsus complicates matters further when he suggests that each organ has a *samen*. He does not elaborate on the idea.
27. *Paracelsus Sämtliche Werke I*, p. 262.
28. Ibid., p. 269.
29. Ibid., p. 270
30. Supporting his argument for the development of children that are part animal, Paracelsus reminds his readers that one frequently hears about such things, 'und solcher hendel sind vil, die alle unter den vorbemelten erzelt werden' (*Paracelsus Sämtliche Werke I*, p. 270). His comment reflects the discussion of unusual births that was common in his day.
31. He says here (*Paracelsus Sämtliche Werke I*, p. 271) that he will write more about the soul in PARAMIRIS, which may be *Volumen Paramirum*. Indeed as demonstrated earlier in this chapter, in that tract he does focus on the soul. This may confirm that *Buch von der Gebärung* may be his earliest surviving tract.
32. 'Und wo die anziehende kraft der matricis nit da ist, da wechset kein geburt. Dan an das ort kompt kein samen, wol ein sperma. Darumb wir nachgedenken den unwissenden arzten, die auf dem spermal ligent als in der natur und wollent die frauen fruchtbar machen und die man, das alles ein erlogene tiedung ist. Wan dieweil die ziehend kraft der matricis nicht do ist aus eigner naur, dieweil kompt der same night an die stat, da wird dan kein kind. Und so der sperma der same were, als sie sagen, der one die muter zu einem kint

wird, wie vormals gemelt ist, so merken das die muter so oft mit im angefüllet würd das alle tag kinder wüchsen. Aber ir lügent ist nicht wol glossirt von inen' (*Paracelsus Sämtliche Werke I*, p. 262).

33. *Paracelsus Sämtliche Werke I*, p. 272.
34. Ibid., p. 272.
35. Ibid., p. 273.
36. Ibid., pp. 287–306. All translations are my own. They are rough translations, intended to give the reader a sense of how Paracelsus wrote.
37. I include the reference in its entirety to demonstrate the difficulty in working through Paracelsus's theory
38. *Paracelsus Sämtliche Werke I*, p. 290
39. Ibid., p. 290.
40. As in the first sentence of the above quote, 'Also frei und ledig one allen zwang hat hat got beschaffen den samen des menschen, das er also still und verborgen im menschen ligt, als das er das selbigen in keinerlei weis noch weg beschwert wird, noch gezwungen' (*Paracelsus Sämtliche Werke I*, p. 290).
41. *'Der man hat keinem samen in im, der do geberen möge seines gleichen, des gleichen die fraun auch nit. Darumb reden unrecht die blinden vermeinten arzt, die da anfüllen große bücher de spermate, dan sie liegen das es sperma sei, das sie dan sperma nennen. Damit wir aber nit unser capitel mit den selbigen blinden partels leuten verfüren und verlengen, wöllen wir euch kürzlich zum text füren unsers fürnemens: dabei ir entlich wissen solt, das gar kein sperma sei, weder in frauen noch in mannen. Auf solchen grunt heben wir unsern text an'.* 'Man has no seed in him, that could bear his own likeness, the same holds true for women. Why then do the dumb doctors talk about it so much, and fill such big books with ideas about sperm. They put forth the idea that it is sperm that is important, and they call it sperm. That is why we write differently and will not fill chapters with dumb explanations about sperm, for there is no sperm-neither in men nor in women. We put forth our text on these grounds' (*Paracelsus Sämtliche Werke I*, p. 291)
42. *'Ihr solt auch gedenken, alles das da gegenwurf werfe seiner beschaffung, das das selbige ist one angangenden samen, welches die unerfarnen arzet sperma nennen, so doch sperme kein samen nit ist, wird auch zu keinem samen'.* 'You should know that all that is planned is his creation, the same is without seed, which the inexperienced doctors call sperm, but sperm is not seed, and will not become seed'. (*Paracelsus Sämtliche Werke I*, p. 292).
43. *Paracelsus Sämtliche Werke I*, p. 295.
44. He uses the term the same way in *Buch von der Gebärung* to mean mind. '*Dise speculation nimpt sich aus dem gegenwurf. Als wan ein man eine hübsche frau sicht, so gibt es ime ein gegenwuwurf und ursach seiner speculation: und so er aus seiner vernunft gedenkt an die ursach irer bescahffung, iezt berhengt er dem willen und mach ein phantasei*' 'This thought is taken from a plan. When a man sees a pretty woman, a plan is given to him and it is the main idea in his mind: and because he rationally thinks about this, his desire makes a fantasy. (*Paracelsus Sämtliche Werke I*, p. 294)
45. *Paracelsus Sämtliche Werke I*, p. 294.
46. Ibid., p.294.
47. Ibid., p. 295.
48. Ibid., p. 296.
49. Ibid., p. 297.
50. *Paracelsus Sämtliche Werke I*, p. 297.

51. This distinction between divinity and nature is more common in his late works as we shall see in chapter 4.
52. *Paracelsus Sämtliche Werke I*, p. 298
53. Ibid., vol. 1, p. xliv.
54. The tract ends abruptly with a list of various concerns about the relationship of divinity to the human body, among other things. The end of the tract reveals Paracelsus's widespread interests. He does not continue his analysis of the nature of the body and of humankind's inner qualities in this text; rather, he turns to explaining the nature of the spirit that can be found in all humans. He seems to rush to the end of his tract, 'so that we can come closer to the end of our task we want to tell you that God gave the internal part of humans all skill, knowledge and reason so that it belongs to the eternal, because it is alive' '*damit wir aber zum ende nahen unsers fürnemens wöllen wir also erzelen, das got dem inwendigen menschen hat geben alle kunst, weisheit und vernunft so in das ewig gehört ist, die weil es beim leben ist*'. (*Paracelsus Sämtliche Werke I*, p. 303). His apparently hastily written conclusion explains the importance of humours and his belief that Adam and Eve had in them all the elements required for life. The seed of Adam and Eve is a part of the seed of all humans and seed is controlled by God the same way that all humans are predestined. Later in his writing, Paracelsus begins to question the idea that humans are direct descendants of Adam and Eve, and that they share the same seed (*samen).* In *Buch von der Gebärung* and De *generation hominis*, Paracelsus signals his conviction that God created the seed that created all humans. Therefore, all humans are directly created by God. It is important to note in the conclusion that he also writes that stars have no bearing on a person's life; they only affect the seed. This conclusion reveals his uncertainty about the forces controlling seed*;* these could be either human imagination, nature, or God. Although he does not arrive at any certainty about the nature of seed in *De generation hominis*, the idea that a person combines exterior (body) and interior (mind, spirit, soul) is clear. While Paracelsus postulates a special connection between humankind and God, in *De generatione hominis* he expresses uncertainty as to whether conception results directly from a divine plan.
55. Crowther, '"Be Fruitful and Multiply"'.
56. G. H. Williams, *Radical Reformation* (Kirksville, MO: Sixteenth Century Journal Publshers, 1992), p. 791.
57. Paracelsus's commentary on the Psalms are undated. Various similar manuscripts exist. Goldammer estimates them to have been written sometime around 1530 or later. The Commentary on the Psalms can be found in *Paracelsus Sämtliche Werke II. Abteilung Band III. Die theologischen und religionsphilosophischen Schriften*, ed. K. Goldammer (Stuttgart: Steiner, 1986) (hereafter Goldammer, II), p. 7.
58. As in the earlier works, he considers the importance of the womb, 'For it is God who makes us in the secrecy of the mother's body and it is Him who knows us' ('*dann got ist, der uns gemacht hat in der heimlichheit mutters leips, und ist, der uns erkennt hat, do wir nocht leimen waren und kot*'). Goldammer, II, p. 16.
59. He seems particularly concerned with the end of Psalm 139 which reads, 'Thou it was who didst fashion my inward parts; thou didst knit me together in my mother's womb. I will praise thee, for thou dost fill me with awe; wonderful thou art, and wonderful thy works. Thou knowest me through and through: my body is no mystery to thee, how I was secretly kneaded into shape and patterned in the depths of the earth. Thou didst see my limbs unformed in the womb, and in they book they are all recorded; day by day

they were fashioned not one of them was late in growing'. *The New English Bible with the Apocrypha Oxford Study Edition* (New York: Oxford University Press, 1970), p. 668.

60. He writes, '*Also werden wir im mutter leip gefangen, in den nieren unser eltern; und unser nieren werden von dir auch gefangen, also daß unser kinder gefangen werden sein wie wir und unser eltern. Darum so mugen wir nit endrinnen aus deiner hand. Dann sie ob uns ist tag und nacht, und so wir in mutter leip ligen so heimlich verborgen, daß auch unser vatter und mutter noch nit wissen haben, so unser gebein, unser substanz noch nit gemacht ist, – so weist du uns. Dann do gibt David zu verstohn, daß ein mensch heimlicher nit kann verboregen sein dann in seiner entpfengnus, so weist sein mutter, sein vatter nix von ihm. Also hat es got geornet, daß nicht auf einmal die empfengnus geschicht, nit zu bestimpter zeit, das ist: so wir am wenigisten dran denken, so ist es do*. Therefore we are created in our mother's body, in the kidneys of our parents, and our kidneys are also from them, so that our children will be conceived the way we were conceived. For we are there day and night, in our mothers body, secretive and hidddien, so that our father and mother do not know about us, we are there before our bones our material are made – you know us. For this is the way David received a son, so that a human cannot be more hidden than within the body of its mother, his father knows nothing of him. And God arranged it this way. (Goldammer, II, pp. 14–15)
61. Goldammer, II, p. 15
62. A translation of a portion of this tract appears in W. B. Deichmann, D. Henschler, B. Holmstedt and G. Keil, 'What is there that is not Poison?: A Study of Third Defense by Paracelsus', *Archives of Toxicology*, 58 (1986), pp. 207–13. The article also describes in greater detail how Paracelsus's treatment of syphilis greatly diverged from the common practices of the day.
63. See: J. Arrizabalaga and K. Siena (eds), 'Medical Responses to the 'French Diesease' in Europe at the turn of the Sixteenth Century', in *Sins of the Flesh: Responding to Sexual Disease in Early Modern Europe* (Toronto: Center for Reformation and Renaissance Studies, 2005). At one point, Paracelsus does echo the sentiments of Galenic physicians when he writes about how it is possible that babies can get the disease by nursing, '...die mutter in französischer vergiftung das kind mit vergifter milch im leib und ausserhalb füret' 'a mother with French disease will nurse the baby with poisoned milk' (*Paracelsus Sämtliche Werke I*, p. 190).
64. R. French and J. Arrizabalaga, 'Coping with the French Disease: University Practitioner's strategies and Tactics in the Transition from the Fifteenth to the Sixteenth Century', in R. French, J. Arrizabalaga, A. Cunningham and L. Garcia-Ballester (eds), *Medicine from the Black Death to the French Disease* (Brookfield, VA: Ashgate, 1998), pp. 248–87. The authors synthesize the learned and lay approaches to the disease in the late fifteenth century. They note that the name 'French Disease' originated in the 1490s when, under the rule of Charles VIII, France invaded Italy, leaving behind the disease (p. 250). Also known as the 'French Pox' a cure for the disease was sought by Cardinal Matthew Lang of Germany who sent court physicians Nicolaus Pol (1470–*c.* 1532) and Paul Ricci (fl. 1512–1532) to Spain to learn about the use of guaiac wood as treatment. Use of guaiac wood in Germany began in 1516 (pp. 257–8). Paracelsus was opposed to the use of guaiac. See: W. Pagel, *Paracelsus*, p. 24.
65. His track on syphilis appears in Sudhoff vol I, pp. 184–366. Webster argues that he was capitalizing on his reputation as a prophet and that he understood how to manipulate the media by focusing on an issue of great concern at the time. See: C. Webster, 'Medicine as Popular Protest in Medicine and the Reformation'. O. P. Grell and A. Cunningham

(eds) *Medicine and the Reformation* (London: Routledge, 1993), pp. 57–77, especially pp. 62–4.

66. *Paracelsus Sämtliche Werke I*, p.206
67. I have argued this point elsewhere. See: A. Cislo, 'Paracelsus's Conception of Seeds: Rethinking Paracelsus's Ideas of Body and Matter', *Ambix*, 55:3 (November 2008), pp. 274–82.
68. He explicitly states his belief that one must reflect on Creation to be the kind of doctor that God intended, 'Von anbegin, das ist von anfang der welt, ist das exempel der arznei gesezt worden, nach welchem wir arzet uns richten sollent' (*Paracelsus Sämtliche Werke I*, p. 270).
69. '*ir wissent das bei den menschen, das die vorgemelten krankheiten alle im leib iren ursprung nemmen, wie dan von inen geschriben stehet. Nun ligen sie im menschen lange zeit verborgen, vor dem ehe sie ausbrechen*' 'You know that in humans all the formentioned illness originiate in the body, as shall be described. They are hidden in humans for a long time before an outbreak' (*Paracelsus Sämtliche Werke I*, p. 199)
70. The original quote reads: '*so du nun bis corpus erkennest so weißt du die wurzeln der krankheit; alsdan kanst du dieselbig heilen*' 'as soon as you recognize the body, you know the root of the illness, and only then can you begin to heal it' (Sudhof I, p.201). Although doctors began to think of infectious diseases as entities, we do not have evidence to suggest that other physicians considered the disease to be a small body. French and Arrizabalaga summarize the prevailing notions of contagion in the following, 'It was not difficult to imagine that the 'matter of disease' which was both evacuated from the body in the living pustules and found in the dead body where the pains had been, was related to or was the same as the 'substance' of the disease. A third correspondence was sometimes also found, that the matter of disease, its substance or entity, was the means of infection. Some writers treat the matter as a sort of yeast, passing from body to body and converting their substance into its own. Others used the image of seeds, so that the disease was like a plant, coming to maturity and spreading itself by means of forms of its own substance. Others again, like Falloppio, took the Galenic doctrine of 'whole substance' and applied it to the disease and also to the body in its reaction to the disease' French, and Arrizabalaga, 'Coping with the French Disease', p. 278.
71. Throughout his writing Paracelsus seeks to understand some aspects of the body as a microcosm. At times the body itself is a microcosm, the womb is a microcosm and, as we have seen, the *liqoure vitae* in *Buch von der Gebärung* is a microcosm of the human.
72. '...der same, daraus das kind wachst, in im hat seiner eltern geftikeit und gesundheit; wie sich aber dieselben eröfnen, das merkent. Alle krankheit so wir haben, die wir uns nicht selbst zufügen, erben wir von der coniuncion unserer eltern. Und ob etwan ein kind krank würd einer andern krankheit, dan bedeutlich seine eltern gehabt heten, so kompt dieselbig aus vermischung zweierlei durcheinander oder bricht im kind aus, das im alten verzogen ist worden' (*Paracelsus Sämtliche Werke I*, p. 204).
73. 'Am ersten zu beschreiben dis buch, ist notwendig zu probiren also, das alle krankheit, so die französischen corpora geben, wonen, ligen und seient in der sperma, und also weiter aus der sperma, so sich die in coitu volkomen gemacht hat, demselbigen in dem er ist, auch den erben zugeteilt wird, und nachfolgents die dasselbig corpus in helt, sich in der anatomei der krankheiten ausspargit und in dieselbig eigenschaft, stett und örter' (*Paracelsus Sämtliche Werke I*, p. 208).
74. 'Darumb nun so der himel den coitum reirt und sein wirkung in der sperma als ein feuer im holz verbringt, so wissent, das er die erbschaft, so vatter und mutter einbilden, dan

verhengt, also das oft ein kin in seins vatters oder mutter art schlecht, solches alles gebürt dem himel zu. Nun verhengt ers an dem ort; dan ursach sie habens auch vom himel und ist ein himlische erbung, welchem der himel nicht zuwider ist. Aus dem folgt nun das der himel väterliche und mütterliche krankheit, so dem himel zustent, erblich last fallen in den samen. Also was krankheit des himels sind, haben ir exaltation in der sperma und entspringent in ir' (Sudhof I, page 212).

75. The original quote reads, '...dan aus dem himel get nichts unempfindlichs, sonder wie der stral also auch sein effectum hat' (*Paracelsus Sämtliche Werke I*, p. 252).
76. 'also dermaßen kocht die Venus in coitu den sperma und überhiziget denselben, also das die krankheiten, so darin ligen, am selbigen ort in solcher gleichnus versotten werden verbrüet' (*Paracelsus Sämtliche Werke I*, p. 213).
77. Quoted from Sudhof I, p. 297. Here, and throughout the tract, he also implies that he considers coitus a wrong or evil act as suggested in the use of the term *spiritus peccans*. This was not an uncommon thought and was seriously considered as early as 1502 (French and Arrizabalaga, 'Coping with the French Disease', p. 250).
78. See Bynum, *Fragmentation and Redemption*, p. 114.
79. Laqueur, *Making Sex*, p. 63.

3 What about Mary? Contemplating Divine and Human Birth

1. See Biegger, '*De Invocatione Beate Mariae Virginis*', pp. 78–87 and Williams, *Radical Reformation*, especially pp. 488–501. Charles Webster helps situate Paracelsus's theological writing by explaining Paracelsus's contacts with other writers, such as Hans Sachs, and Paracelsus's opposition to the Church in *Paracelsus* especially pp. 70–98. For more on the French and Italian discussions of Mary from the medieval to the early modern period see: D. S. Ellington, *From Sacred Body to Angelic Soul* (Washington, DC: Catholic University of America Press, 2001). Ellington argues that perceptions of Mary changed in the sixteenth century when sermons tended to focus on the spiritual qualities of Mary, rather than on her body.
2. Biegger, '*De Invocatione Beate Mariae*', p. 72.
3. Erasmus of Rotterdam rejected what he viewed as an overemphasis on Marian worship calling instead for the recognition of salvation through Christ. Reformers including Luther, Zwingli and Calvin rejected most aspects of devotion to Mary. Paracelsus was clearly familiar with discussion concerning the significance of Mary in Christian thought and interjected his own ideas on the subject as K. Biegger and U. Gause have shown.
4. Andrew Weeks explains how Paracelsus's writing on Mary in the early years demonstrates his adherence to doctrinal views and forced him to flee Salzburg, where he had been harshly criticized for his viewpoint in *Paracelsus: Speculative Theory*, pp. 79–83.
5. What I mean here is that Paracelsus contemplated the realness of Mary's body as a manifestation of God in a similar way that late medieval German theologians contemplated the presence of God in the Eucharist. See C. Elwood, *The Body Broken* (New York: Oxford University Press, 1999) and C. W. Bynum, *Wonderful Blood: Theology and Practice in Late Medieval Germany and Beyond* (Philadelphia, PA: University of Pennsylvania Press, 2007), pp. 77–8.
6. Biegger, '*De Invocatione Beatae Mariae Virginis*', pp. 12–13. Biegger also provided foundation for future research with her detailed assessment of available religious tracts. She includes remarks by Paracelsus's contemporaries regarding his theological interests. Heinrich Bullinger (1504–75) wrote that Paracelsus was more interested in magic than

religion and spent too much time pondering the significance of demons. By contrast, Johanes Huser, the first compiler of Paracelsus's works, wrote of Paracelsus's devotion to the Catholic Church. Bullinger wrote of conversations he had with Paracelsus, whereas Huser knew Paracelsus only through the texts he acquired after Paracelsus's death. Yet it should be kept in mind that Bullinger is a known anti-paracelsian. Paracelsus may have been more outspoken about demonic forces than he considered in his writing, or it is possible that some texts did not survive. Biegger describes Paracelsus as neither Catholic nor Protestant and someone who was often considered a heretic by his peers.

7. Biegger argues that his perception of God is based on the Old Testament and therefore is personal and patriarchal. She writes, 'Einfältig, schlicht und gläubig entnimmt Paracelsus seine Vorstellung der Bibel; sein Gottesbild ist alttestamentlich personal, patriarchal. Gott als Schöpfer der Welt, Vater des Menschens und Urheber allen Lebens steht für den Arzt und Naturforscher im Vordergrund. Im naturphilosophischen und kosmologischen Teil von Hohenheims Werk finden sich zahlreiche Überlegungen, die den Schöpfungsbericht der Genesis umschreiben und konkretisieren', Biegger, *'De Invocatione Beatae Marae Virginis'*, p. 51.
8. Biegger explains that describing the heavenly and earthly body is a theme throughout his entire oeuvre, 'Die komplexen, ineinander verschachtelten Verhältnisse zwischen materiaellen und geistig-seelischen Anteil des Menschen, zwischen Natürlichem und Göttlichem, zwischen irdischer und himmlischer Leiblichkeit haben Hohenheim Aufmerksamkeit stets erregt und gefesselt; die Faszination dieses Themas liess ihn, quer durch sein ganzes Werk, nicht los' *De Invocation e Beatae Mariae Virginis*, pp. 51–2.
9. Gause, *Paracelsus*. Gause recognized the strong influence of Paracelsus's medical interests in his interpretation of Christ and Mary.
10. In his dissertation, Dane Thor Daniel investigated Paracelsus's biblical interpretation as a means of defining the theological interpretation of the eternal body in Paracelsus's *De Geneologia Christi.* Daniel focused on the meaning of baptism in several of Paracelsus's theological works and related Paracelsus's writing to biblical passages in an attempt to uncover which passages most influenced Paracelsus's anthropology. He tended to focus on the development of a spiritual body in Paracelsus's writing and his work helps articulate Paracelsus's theology. D. T. Daniel, 'Paracelsus on Baptism and The Acquiring of the Eternal Body', in *Paracelsian Moments*, ed. G. S. Williams and C. D. Gunnoe, Jr (Kirksville: Truman State University, 2002), pp. 117–34 and 'Invisible Wombs: Rethinking Paracelsus's Concept of Body and Matter', *Ambix*, 53 (2006), pp. 129–42.
11. *Paracelsus Sämtliche Werke II*, vol. 3, pp. 235–66.
12. Relating understanding of human birth to Christ's was not uncommon in the sixteenth century. Williams, *Radical Reformation*, noted that Reformers' interpretation of the birth of Christ often relied on either an Aristotelian model of conception or a Lucretian-Hippocratic view that the male and female were equally contributive in the generation of progeny. He describes four Reformers thusly, '[Clement] Ziegler ... believed that Christ brought his translucent body with him from heaven and acquired visibility form the flesh born of Mary. [Melchior] Hofmann postulated a single, divine nature from heaven but called it celestial flesh...[Michael] Servetus, though he originally distinguished between the prolation of the Word and the generation of the Son, came to speak of three phases of generation and granted that a Christ of two natures took one of them from Mary, who was therefore truly Theotokos. Schwenckfeld postulated two natures, the one celestial, of God the Father and God the eternal Son, and the other human but "uncreaturely"' Williams, *Radical Reformation* p. 490.

13. *Paracelsus Sämtliche Werke II*, vol. 3, p. 238
14. Alan Debus research demonstrated how these three core principles formed the basis of Paracelsus's chemical philosophy and they in turn were elements of a particular approach to Alchemy that resonated in the works of several French followers. See: Debus, *The Chemical Philosophy*.
15. I am referring to Bruce Moran's chapter on Paracelsus and Paracelsians in his book, *Distilling Knowledge*, pp. 67–98.
16. In some of the manuscripts this word was 'im' see footnote 29 in *Paracelsus Sämtliche Werke II*, vol. 3, p.240.
17. *Sämtliche Werke, II*, vol. 3, p. 240.
18. Gause, *Paracelsus*, p. 13.
19. *Sämtliche Werke, II*, vol. 3, p. 240.
20. *Paracelsus Sämtliche Werke II*, vol. 3, p. 259.
21. Explorations of the significance of Mary and her relationship to God were common during the Reformation, however most mainstream Reformers regarded Mary's role differently than Paracelsus. In 1523 Carlstadt wrote in his *Fürbit Mary*, (*On the Intercession of Mary*) 'I would gladly honor Mary, but I do not want to deprive her Creator [God the Father] and Lord [Christ] of their honor, and attribute to Mary that which she would repel' quoted from Calven A. Pater, 'Lay Religion in the Program of Andreas Bodenstein von Karlstadt', Leaders of the Reformation, ed. Richard L DeMolen (Selinsgrove, Pa.: Susquehanna University Press, 1984), p. 71 quoted in Williams, *Radical Reformation* (Kirksville, MO: Sixteenth Century Studies, 1992) 480. See also: Jarsolav Pelikan, *Mary Through the Centuries: Her Place in the History of Culture* (New Haven, CT: Yale University Press, 1996).
22. Quoted from *Paracelsus Sämtliche Werke II*, vol. 3, p. 245. Ute Gause regards the relationship of Mary to God in *Liber Sancta Theotica* as representative of Paracelsus's conviction that the male and female compliment and complete each other. She writes, 'the femininity in the Goddess Mary is the complementary side of the manliness of God the Father. She is her own property, without her God is unthinkable for Paracelsus' '*Die Weiblichkeit Gottes in Maria ist die komplementäre Seite zur Männlichkeit Gottes, des Schöpfers. Sie ist eine Eigenschaft, ohne die Gott für Paracelsus undenkbar ist*' Gause, *Paracelsus*, p. 14.
23. 'For there is an additional difference that must be recognized: God the sun is a more powerful God in his being. But the heavenly woman has no power in her being'. '*Da ist aber weiter ein undterscheid zu merken: gott der sun ist ein gewaltiger gott in seiner person. Aber die himblische frau nit, hat nichts in ir person zu gewaltigen. Gott der heilig geist ist auch gewaltig in seiner person und ein gewaltiger gott, aber die himblisch frau nicht*', *Sämtliche Werke*, vol. 2, 3, p. 245.
24. Weeks, *Paracelsus: Speculative Theory*, pp. 84–5.
25. L. Braun, 'L'Idée de ‚Matrix' chez Paracelse' Heinz Schott and Ilana Ziguer (eds) *Paracelsus und Seine Internationale Rezeption in der frühen Neuzeit* (Leiden: Brill, 1998) suggested that Paracelsus scholarship investigate the way Paracelsus describes Mary's role in the trinity as a way to understand the idea of matrix within Paracelsus's oeuvre.
26. Though including Mary in the Trinity was not common among radical reformers, some of the Antitrinitarians, most of whose writings stem from the period after 1530, explore the relationship between Christ's body and God. Gause interprets Paracelsus's constellation of Mary in the Trinity as Paracelsus's attempt to build on his often used framework of men and women as complementary opposites, 'Die Weiblichkeit Gottes in Maria ist die komplementäre Seite zur Männlichkeit Gottes, des Schöpfers. Sie ist eine

Eigenschaft, ohne die Gott für Paracelsus undenkbar ist. Diese Komplementarität gehört essentiell zum Gottesbild, so daß Maria vollkommen integriert ist', Gause, *Paracelsus*, p. 14.

27. Quoted from *Paracelsus Sämtliche Werke II*, vol. 3, p. 246. The notion of Mary as a queen was common throughout the medieval period. For medieval preachers, Mary's ability to participate in human redemption stemmed from her divine motherhood of Christ and that Mother and Son possessed a common flesh and humanity. See: Ellington, *From Sacred Body to Angelic Soul*, p. 48.
28. *Paracelsus Sämtliche Werke II*, vol. 3, p. 246.
29. He writes, 'while at the same time they are born of the queen and are divided within their being and also complete through this birth, for that reason she uses her will through the beings and not through her own being' '*wann zugleicherweis wie sie durch die künigin geboren seindt und in der person geteilt und durch sie also geporen und vollendet, also herwiderumb verbringt sie iren willen durch die personen und nicht durch ir eigen person*' *Paracelsus Sämtliche Werke II*, 3, p. 247. Paracelsus's wish to distance Mary from other humans is detailed in his tract *Libellus de Birgine Sancta Theotica* (1524). The tract is not included in the collected works, but Ute Gause examined the extant manuscripts and explored the birth of Mary in them. She notes that Paracelsus explained that Mary, 'is not like any other human born woman, born from mortal seed and the way our mother was born' '*ist nicht 'gleich einer andern geborenen frauen aus tödlichem samen und wie unsere müetter geborn sind*' (178b)' (Gause, *Paraclelsus*, p. 27).
30. Quoted from *Paracelsus Sämtliche Werke*, vol. 3, p. 247. The '*paramirishen*' tracts could either be the *Volumen Paramirum* or the *Opus Paramirum*. Both have been previously categorized as part of his medical writing. These tracts shall be examined in Chapter 4.
31. In Biegger, *'De Invocatione Beate Marie Virginis'*, pp. 150–88.
32. 'There is a difference between the creatur (s) and the holy. The holy come directly from God. The creatur(s) emerge from . For that reason we gravitate toward not toward the creatur(s) but rather to the holy ones...we would not recognize these without God' '*es ist ein unterschaidt zwischen der creatur und den heiligen. Der hailige gehet von Gott auß. Die creatur gehet von geschöpff aus. Darumb so lauffen wir nit zur creatur, sondern zum hailligen, über den Gott zeugkhnuß gibt, das wir ohn irrsahl dieselben auß Gott wohl mügen erkhennen*' Paracelsus, '*De Invocatione Beate Marie Virginis*' in Biegger, *'De Invocatione Beate Marie Virginis'*, p. 158.
33. Biegger, *'De Invocatione Beate Marie Virginis'* pp. 159–60. Biegger notes that this interpretation of Mary differs from what Paracelsus writes in *De Geologia Christi*, in which he wrote, 'Mary was the first of all creatures, the first born and her son was the first born among men' '*Maria ist die aller erst undter allen creaturen gewesen, die erst geboren, und ir son der erst geboren, und ir son der erst geboren undter den mannen*', *Sämptliche Werke*, vol. 2, 3, p. 139.
34. Biegger, *'De Invocatione Beatae Mariae Virginis'*, p. 160
35. 'Eve comes from the earth and returns to it. Maria came directly from God' 'Eua geht auß der erdten und wider darein. Maria geht von Gott aus'. Biegger, *'De Invocatione Beatae Mariae Virginis*', p. 160. Eve came from the earth and went back to it. Maria came from God. This was a common understanding of Mary in medieval sermons. See: Ellington, *From Sacred Body to Angelic Soul*.
36. The quote continues, 'ist das nit sovil: den ich trage, der ist ewer heillandt. Drumb ist all ewer trosst in mir'. Biegger, *'De Invocation Beatae Marie Virginis'*, p. 164.
37. Ibid., pp. 164–5

38. Katharina Biegger identifies a possible motivation for Paracleus' fascination with Mary: 'Als tiefliegende Motive seiner Beschäftigung mit der Gottesmutter kann man wohl ebenfalls Frangen der Körperlichkeit in Zeugung, Geburt, Tod und Auferstehen vermuten' in 'Paracelsus über Maria-Weiblichkeit und Heiligkeit' *Nova Acta Paracelsica*, 14 (2000), p. 6. Biegger reads *De Invocatione Beatae Mariae Virginis* as one of Parcelsus's most reserved writings on Mary. She notes that it was one of the last he wrote on the subject. In earlier writing (1524), such as *Liber de sancta trinitate*, he was more outspoken about Mary's role as a Goddess in the Trinity and her role as a helpmate to God.
39. Biegger, *'De Invocationen Beatae Mariae Virginis'*, p. 166
40. Ibid., p. 184
41. *Paracelsus Sämtliche Werke II*, vol. 3, pp. 59–164
42. Ibid., p. xxxi.
43. Ibid., pp. 63–6.
44. He writes, 'die nit aus gott geboren seindt, denselbigen ist kein incorporation gegen der trinitet, noch der trinitet mit ihnen, zu erkennen, sonder geschieden das bös vom guten, das finster vom liecht. es incorporirt sich nichts mit der trinitet denn das, so aus gott geboren ist und wird' *Paracelsus Sämtliche Werke II*, vol. 3, p. 75
45. Paracelsus, *De Genealogia Christi*, in *Paracelsus Sämtliche Werke II*, vol. 3 (1986), p. 123.
46. *De Genealogia Christi*, pp. 124–7
47. Ibid., p. 128.
48. Dane Thor Daniel interpreted this tract as an example of Paracelsus's very literal reading of the Bible arguing that Paracelsus employed his natural philosophy to the birth of Christ. I disagree and see the need to read the tract in context of Paracelsus's other writing. When considered in context, it becomes clear that Paracelsus's contemplation of Christ actually informs his natural philosophy. For Daniel's interpretation see: 'Paracelsus on Baptism and the Acquiring of the Eternal Body', in G. S. Williams and C. D. Gunnoe (eds.) *Paracelsian Moments*. (Kirksville: Truman State University Press, 2002), pp. 117–33.
49. 'Notice therefore that the corporeality of Chirst can be compared to the corporeality of the Holy Ghost: therefore we have seen the heavenly Spirit in heaven with the same corporeality, that I mention' '*also auch merkent die corporalitet Christi in vergleichung der corporalitet des heiligen geists: also daß wir den heiligen geist himblisch gesehen haben in derselbigen corporalitet, wie dan gemelt ist*', *Paracelsus Sämtliche Werke II*, vol. 3, pp. 61–2.
50. 'I think that it is a good idea to understand the third person in the trinity because I describe the lineage of the, birth of Christ, so that we more thoroughly understand Christ as part of Godliness as the son of God, as the other Adam, as the first born from God, from whom we are all in the new birth children of God, in blood and flesh and in our bodies, as the heavenly body wishes it. *Also hat mich für gut angesehen, daß ich die dritt person der trinitet von wegen der genealogia Christi und der neuen gepurt beschreibe, auf daß ir dester grundlicher Christum in der gottheit erkennent als den son gottes, als den andern Adam, als den erstgebornen aus got, von dem wir in der neuen gepurt seindt auch kinder gottes, in blut und fleisch udn des leibes, so die himlisch corporalitet vermag'*, *Paracelsus Sämtliche Werke II*, vol. 3, pp. 62–3
51. 'For in the natural the supernatural will be understood in the natural, that is in humans' '*dann aus dem natürlichen werden die übernatürlichen verstanden in dem natürlichen, das ist in dem menschen*', *Paracelsus Sämtliche Werke II*, vol. 3, p. 64.

52. 'It is, that the nature is like the number and the workings of the trinity – except nature has less of a creator. Nevertheless, everything was ordered in His plan following the order of the monarchy of godly existence, in the same way that the image of human comes from God', '*das ist, das die natur gleich ist in der zal und operation wie die trinitet – als allein, sovil die natur hat weniger als ein schöpfer. iedoch alles nach ordnung seines geschöpfs gemacht nach der ordnung und monarchei göttlicher eigenschaft, zugleicherweis wie die bildnus des menschen von gott gehet*' (*Paracelsus Sämtliche Werke II*, vol. 3, p. 65). The construct of regarding nature as reflective of the trinity calls to mind his well-known chemical theory the three primary substances: mercury, sulphur and salt.
53. 'Concerning the Son, he is the other person of the trinity, so that a new creatur was created, a new human, one that is eternal and has an immortal body. '*Sovil aber und als ein son betrifft, das ist die ander person der trinitet, ist das sein gab, daß er ein neue creatur geschaffen hat, ein neuen menschen eins ewigen, untödlichen leibs. Paracelsus Sämtliche Werke II*, vol. 3, p. 71.
54. *De Genelogia Christi*, p. 76.
55. 'Another thing that I must call your attention to is this: all things return to that from which they came. Therefore Adam was from the earth and will return to it' '*Aber ein undterricht ist zu merken; nämblich in dem, daß alle ding wiederumb komben in das, aus dem sie komben seindt. Also ist Adam von der erden gemacht worden und wiederumb in dieselbig komben*', *Paracelsus Sämtliche Werke II*, 3 pp. 77–8.
56. *De Geneologia Christi*, p. 76
57. Ibid., p. 82.
58. *Paracelsus Sämtliche Werke II*, vol. 3, pp. 95–6.
59. 'So the bodily body of Christ, that is mortal, was not earthly, it was not material like it is in humans'. '*Also ist der leiblich leib Christi, der tödlich gewesen ist, nit irdisch leiblich gewesen, nit materialisch als im menschen ...*' (*Paracelsus Sämtliche Werke II*, vol. 3, p. 96)
60. Ibid., p. 97.
61. 'Even though the body decays and the spirit is with the Father, where it belongs, he is still not completely dead. They both must come together for the judgement of God. Only then will death be complete and remain with the damned and give way to the living for salvation, as he is guided through Christ in the resurrection' '*ob gleichwol der körper faulet und der geist beim herren ist, wo er hin gehört, noch so ist er nit vokomben gestorben. Sie beide muessen wieder zusamben und für das gericht gottes. Alsdann so wird der tod volkomben werden und bleiben bei den verdambten und weichen von den lebentigen zur seligkeit, wie er gewichen ist von Christo in der auffersteheung*' (*Paracelsus Sämtliche Werke II*, vol. 3, p. 97). He goes so far as to believe that on the third day after death humans will ascend, ' On the third day is our resurrection from the grave from where we are buried' '*in der dritten zal wird auch stehen unser auferstehung von den gräbern, und wa wir vergraben ligent*' (*Paracelsus Sämtliche Werke II*, vol. 3, p. 98).
62. Goldammer II, vol. 3, p. 130
63. His understanding of predestination agrees with Calvin's belief that God actively wills the salvation of those who will be saved and the belief that the quality of being eternal is predestined by God. However thinking about predestination and salvation in this way did not gain momentum in the Reformation until around 1570, well after Paracelsus's death. A. McGrath, *Reformation Thought* (Oxford: B. Blackwell, 1988), pp. 91–2.
64. *Paracelsus Sämtliche Werke II*, vol. 3, p. 139.

65. Ibid., 3, p. 216 The way Paracelsus defines Mary's birth here agrees with his belief that there is a close connection between Mary and God as he wrote in *Liber de sancta trinitate.*
66. Biegger, *'De Invocatione Beatae Mariae Virginis'*, p. 216.
67. *Paracelsus Sämtliche Werke II*, vol 3, p. 139.
68. Ibid., p. 142.
69. For a discussion of the elemental spirit in Paracelsus's writing see Williams, *Defining Dominion.*
70. She miraculously carries Christ in her womb, 'darumb in der vile der auserwöhlten wird sie das lob erlangen wunderbarlich über alle wunder, nämblich daß sie die ist, die den schöpfer himmel und erden tragen soll in irem bauch', *Paracelsus Sämtliche Werke II*, vol. 3, p. 151.

4 Adam, Eve and the Human Body: Paracelsus's Nature Dilemma

1. The summary of ideas on the soul is based on Kristeller, *Renaissance Thought*, pp. 181–96.
2. N. Siraisi, *The Clock and the Mirror* (Princeton, NJ: Princeton University Press, 1997), p. 66.
3. Kathleen Crowther pointed out this trend in '"Be Fruitful and Multiply"'. However, in this article her examples all date after Paracelsus's writing on the topic. It is possible that Paracelsus helped popularize the use of Adam and Eve to explain the nature of the soul in tracts on human birth.
4. M. Bergengruen compares Paracelsus's 'alten Adam' to Luther's use of the same trope in *Nachfolge Christi-Nachahmung der Natur* (Hamburg: Meiner, 2007), p. 55ff and Weeks, *Paracelsus: Speculative Theory.*
5. *Paracelsus Sämtliche Werke I*, vol. 1 pp. 163–239 Sudhoff considers *Volumen Paramirum* one of Paracelsus's earliest texts although the dating remains imprecise. The tract seems to have been written sometime around the time he *Liber de Sancta Trinitate* (see Chapter 3), although in that tract he refers to his *paramirische[n] schriften* as forthcoming. Andrew Weeks agreed with Sudhoff that *Volumen Paramirum* may be the earliest known work, but evaluating Paracelsus's understanding of embodiment in this tract suggests that it is indeed a later, or what I prefer to call a transitional work because it bridges the philosophical divide between Paracelsus's comparison of the birth of all bodies to Christ's and his later delineation of natural and spiritual bodies. Weeks, *Paracelsus: Speculative Theory*, pp. 38 and 50.
6. He rejects his earlier claim human birth is divinely willed. His expression of human spirituality in this tract aligns with sentiments expressed in *De geneologica Christi.* Recalling that *De geneologia Christi* also dealt with the degree to which the propensity to be spiritual was part of a human generative process, I wish to consider *Volumen Paramirum* a similar development in Paracelsus's understanding of the dual nature of humans. He writes, 'Die paramirischen schriften seindt uns wol zu betrachten in dem, das das liecht der natur und der heilig geist uns anzeigt. Dann von ihm gehts alles aus in unsern gedanken' (*Paracelsus Sämtliche Werke I*, vol. 1 241). He uses the term *liecht der natur* in as the essential link *volumen paramirum* to knowledge and says that all physicians must be attuned to the light of nature. *Volumen Paramirum* implies that Paracelsus called for physicians to be Christian and recognize that the Holy Spirit guides them in uncovering the secrets of God's plan.

7. *Paracelsus Sämtliche Werke I*, vol. 1, pp. 167–8. He also repeatedly directs his remarks at what he believes are 'misguided' physicians such as those who are not attuned to nature. He names Hippocrates and Galen as learned men who disregard nature (*Paracelsus Sämtliche Werke I*, vol. 1, p. 170).
8. *Paracelsus Sämtliche Werke I*, pp. 175–6.
9. The idea of corporality, specifically the use of the term *corpora,* appears infrequently in his works prior to 1537. Another instance where it occurs is *liber sancta theotica*. The quoted passage can be found in *Paracelsus Sämtliche Werke I*, pp. 176–7.
10. 'Adam and Even received their body through the creation and through the *ens seminis* until the end of the world' '*Adam und Eva haben ire leib durch das geschöpf empfangen und durch ens seminis bis in zergehung der welt*,' *Paracelsus Sämtliche Werke I*, p. 177.
11. *Paracelsus Sämtliche Werke I*, vol. 1 p. 237.
12. Reiterating his stance that only Adam was created by God he writes, 'Life was divinely given to Adam and his seed given further, rather everything that nature forms from her ens, the same she gives to life, for it was given only once, from which comes growth, strength, and all the other things a human must have in him' '*das leben ist götlich geben worden Adae und sein samen und wirt weiter ninen genomen, sonder alles das, das die natur formirt aus irem ens, dem selbigen hat sie das leben auch zugeben, dan einmal ist es geben worden, wird weiter nimen geben, aus dem kompt das wachsen, die sterk, die andern so der mensch an im hat und haben muß*' *Paracelsus Sämtliche Werke I* , vol. 1, p. 237
13. *Volumen Paramirum*, *Paracelsus Sämtliche Werke I*, vol. 1, p. 179
14. Employing the model of microcosm and macrocosm to God's creation as a way of understanding the natural world was common as early as the medieval period and other have noted that this paradigm led to modern science. P. Harrison, *The Bible, Protestantism, and the Rise of Natural Sci*ence (Cambridge: Cambridge University Press, 1998) explains, 'Whereas micorcosm-macrocosm had enabled exegetes to establish the meaning of other living things, with creatures in the external world being interpreted as features of the inner spiritual and moral world, now this same principle posited man as a material being, embedded in a material world, and intimately connected with the whole creation' (p. 50) It should be noted that in *Volumen Paramirum* Paracelsus still distinguishes between humans and the other things of the material world that lack spiritual ties to God.
15. *Paracelsus Sämtliche Werke I*, vol. 1, p. 218.
16. Paracelsus's idea that receiving the spirit was a conscious act coincides with the logic some of the radical reformers used to explain the importance of re-baptism,that is adult baptism.
17. Quoted from *Paracelsus Sämtliche Werke, I*, vol. 1, p. 218. For more on how Paracelsus works with the idea of human embodiment and mental processes see: H. C. E. Midelfort, 'The Anthropological Roots of Paracelsus's Psychiatry' *Medizinhistorisches Journal*, 16 (1981), pp. 67–77. Midelfort explains the formation of Paracelsus's 'Psychiatry' in regard to the concept either his bipartite schema of the human or his tripartite schema of humans. On another note, Paracelsus's formulation of how one receives the spirit relegates the young to those lacking this connection to the divine. For if one were to accept Paracelsus's reasoning of how a body receives a spirit, it would be difficult to regard a young child as human at all. For a further discussion of how childhood was understood in this period see: K. Arnold. *Kind und Gesellschaft in Mittelalter und Renaissance*. Beiträge und Texte zur Geschichte der Kindheit, Paderborn 1980 and Dieter Richter. *Das fremde Kind: Zur Enstehung der Kindheitsbilder des bürgerlichen Zeitalters*. 1987.

18. Will-Erich Peuckert, a modern compiler of Paracelsus's works, thought that *Astronomia Magna* was the most representative of Paracelsus's oeuvre and chose it to introduce readers to Paracelsus in his edition of Paracelsus's collected works. The assumption that *Astronomia Magna* represents the pinnacle of Paracelsus's achievement is widespread in Paracelsus scholarship. Although Paracelsus reaches clarity on differentiating qualities of the human body and soul in this tract, using *Astronomia Magna* as a key to all of Paracelsus's other writing confuses his influence in the field of medicine. Indeed, the way that Paracelsus came to define the human body in this tract does not seem to be the approach to the body that his followers adapted, nor are his interpretations of Adam and Eve consistent with his contemplation of these biblical figures elsewhere in his oeuvre. A recent example of the assumption that Astronomia Magna represents the pinnacle of Paracelsus's achievement can be found in: D.T. Daniel, 'Invisible Wombs'.
19. *Paracelsus Sämtliche Werke I,* vol. 12, pp. 3–4
20. Ibid., p. 9
21. Ibid., p. 9
22. Ibid., p. 12
23. U. Gause, 'On Paracelsus's Epistemology', in O. P. Grell, *Paracelsus: The Man and His Reputation His Ideas and their Transformation* (Leiden: Brill, 1998), p. 216 argues that Paracelsus recognizes that knowledge of God through nature in *Astronomia Magna* is a way of, 'directing the mortal body from going astray'. Gause posits that Paracelsus concerns himself with nature primarily as a way of knowing God.
24. *Paracelsus Sämtliche Werke I,* vol. 12, p. 39
25. Ibid., p. 39.
26. Ibid.
27. Ibid., p. 45.
28. 'For the first human was handmade by God and his woman was born and given to him, thereafter the order of nature took over. That is, a work is ordered in nature, through which a human will never again be made, but will instead be born one from another. For Adam alone was made directly by God'. '*Dan den ersten menschen hat die hant gottes gemacht und im sein frau auch ungeboren geben, weiter ist er der ordnung der natur uberlifert worden. Das ist, ein werk ist geordnet in die natur, durch welche der mensch nicht mer weiter gemacht, sonder geboren sol werden, ie einer von dem andern. Dan alein Adam der was gottes*' (*Paracelsus Sämtliche Werke I,* vol. 12, p. 45)
29. Recall that in previous tracts he insisted on a two-seed theory of conception.
30. *Paracelsus Sämtliche Werke I,* vol. 12, p. 49.
31. See Daston and Park, *Wonders and the Order of Nature* and J. Spinks, *Monstrous Births and Visual Culture in Sixteenth-Century Germany* (London: Pickering and Chatto, 2009).
32. He writes, 'Aber leser, wie du hie in diser vorred verstehn magst unt lisest, das ich die philosophia des innern himel zu beschreiben understehe, durch die hülf gottes: Sost dich nit lassen befremden oder eußerlich hinlassen gehen, darumb das ich schreib, mer theologisch als natürliche philosophei, so ist doch die ursach, das der eußer himel gescheiden ist von dem innern, darumb kein natürlich philosphia mag eingefürt werden, alein was die theologische philosophia vermag' (*Paracelsus Sämtliche Werke I,* vol. 12, p. 283).
33. *Paracelsus Sämtliche Werke I,* vol. 12, p. 287.
34. *Paracelsus Sämtliche Werke I,* vol. 12, p. 288. However it should be noted that Paracelsus believes that Christ was created before Adam. He explains, 'Christ was the first human and he was for Adam, just like He said to the Jews: Before Abraham was there: I was

there' '*...Christus ist der erste mensch gewesen und gewesen für Adam, gleich so wol, als da er gesagt hat zu den Jüden: ehe das Abraham gewesen ist, da bin ich gewesen*' (*Paracelsus Sämtliche Werke I*, vol. 12, p. 288).

35. *Paracelsus Sämtliche Werke I*, vol. 12, p. 290. He reiterates this construct later, 'For the spirit alone is from God, that's why it returns to God, for it was given to him from the heavens, and the body is from the earth and it will return to the earth in dust and ashes' '*nun ist der geist alein von got, darumb so gehet er wider zu got, der in geben hat von himel, und das fleisch ist nun ein erden, aschen und staub, in das wird er wider*' (*Paracelsus Sämtliche Werke I*, vol. 12, p. 295). Following Scripture he writes, 'Wiewol die schrift sagt: in meinem fleisch werde ich sehen meinen got, meinen heilmacher ...' (*Paracelsus Sämtliche Werke I*, vol. 12, p. 295).
36. *Paracelsus Sämtliche Werke I*, vol. 12, p. 297. He also says that there is another kind of spirit that is found in all living things, but that he does not intend to discuss that spirit in this tract, 'nin ist aber ein ander giest, der ist natürlich, der ist im menschen wie im vihe. Und der selbig geist wird hie nicht verstanden'.
37. *Paracelsus Sämtliche Werke I*, vol. 12, p. 298.
38. Ibid., pp. 298–9.
39. At times Paracelsus's interpretation of the birth of Christ appears anachronistic, as when he explains that Christ was the first human, created before Adam, '...Christus ist der erste mensch gewesen und gewesen für Adam...', *Paracelsus Sämtliche Werke I*, vol. 12, p. 288.
40. *Paracelsus Sämtliche Werke I*, vol. 12, pp. 307–8.
41. Ibid.
42. See Chapter 3.
43. The two concepts become intricately tied when he tries to trace the lineage of humans through Christ and Adam to explain Resurrection,

> *Und zu gleicher weis, wie Christus alein von einer jungfrauen geboren ist, der one menlichen samen von Adam mensch ist worden und in ir incarnirt, und mensch worden durch den heiligen geist, also wir menschen, die da wöllen in das ewig leben, das tötlich fleisch und blut verlassen müssen, wider zum andern mal geboren werden von einer jungfrauen und glauben, incarnirt vom heiligen geist: als dan seind wir fehig in das reich gottes. Dan einmal so muß der mensch blut und fleisch sein in ewigkeit. Darumb sind zwei fleisch: von Adam, das ist nichts wert, von dem heiligen geist, das macht lebendig fleisch, dan er incarnirt von oben herab, darum komet sein incarnation durch uns wider gen himel. also wissent nun, das der tauf anstat der jungfrauen das ist und im tauf werden wir incarnirt vom heiligen geist, nemlich von dem heiligen geist...* (*Paracelsus Sämtliche Werke I*, vol. 12, p. 309)

The distinction between the natural body and the spiritual body, so difficult for Paracelsus to identify, is lost when he redefines the two bodies as related to Adam and Christ. He reads Christ's birth as an act to be emulated by the human body seeking Redemption. In other words, Paracelsus associates the divine birth of Christ with the spiritual birth of humans, all the while seeing the need to validate this birth through baptism. Baptism represents a divine gift that is intended to compensate for the absence of the virgin birth. His thoughts on redemption share in Christian beliefs in redemption of the soul, but Paracelsus's conflates this belief with his delineation of the spiritual and corporeal natures of humans. Furthermore he accepts faith as a spiritual body because he understands that only those who are saved through Christ have an immortal body; and some people are predestined to be immortal. In this sense, his understanding of the dual nature of the body is limited to Christians, who, as he explains in this tract, are predestined to believe in Christ.

44. *Paracelsus Sämtliche Werke I*, vol. 12, p. 316
45. 'aus dem folgt nun, das die geschriften nicht sollen interpretiert werden durch die natürlich weishet, noch die natürlichen geschriften durch den geistlichen verstant'(*Paracelsus Sämtliche Werke I*, vol. 12, p. 317).
46. *Paracelsus Sämtliche Werke I*, vol. 12, p. 318.
47. Ibid., vol. 12, p. 331.
48. Acknowledging the possibility of evil that enters the human body, Paracelsus explores and defines his thoughts on magic (*magie*), the supernatural. It has been argued that Paracelsus's concern with nature relates to his professional quest to assert the importance of magic (*Magie*) in medicine. K. Goldammer, *Der Göttliche Magier und die Magierin Natur* (Stuttgart: Franz Steiner, 1991), pp. 45–57. Goldammer understands Paracelsus attempts to define the human body as part of the cosmos part of Paracelsus's hope to legitimate his approach to healing
49. *Paracelsus Sämtliche Werke I*, vol. 9, pp. 177–230
50. Ibid., vol. 9, p. 190.
51. Laqueur, *Making Sex*.
52. Gunhild Pörksen wrote of Paracelsus's interest in women's medicine, defining it as a call for gynocology, in 'Microcosma. Der weibliche Leib. Zu einer Theorie der „Frauenheilkunde" bei Paracelsus'. *NovaActa Paracelsica*, 14 (2000), pp. 79–96.
53. *Paracelsus Sämtliche Werke I*, vol. 9, p. 192
54. Walter Pagel drew a connection between Paracelsus's model of reproduction and Albertus Magnus's constellation of the universe in *Philosophia Pauperum*. Pagel focuses on the woodcut printed with the second edition of Magus's text of 1493 in which the earth is represented as the womb of the cosmos (pp. 238–9).
55. Hildegard Keller considers Paracelsus's emphasis on the 'invisible' nature of the womb part of his philosophy of observation and his gendered understanding of medicine, but she suggests that the invisible nature of the womb 'does not invest them with danger and excitement. Nonetheless, he gives to them and to the whole woman (i.e. both of the worlds she is) a special dignity: they do deserve to be known by women and men', 'Seeing Macramosa', p. 99.
56. 'zu gleicher weis wie er tragen wird auf dem wasser, also tragen wir menschen in auf erden und aufm wasser, welchen geist nie keiner gesehen hat noch gesicht, und ist der der da ist in matrix der menschen, das ist in der frauen. Darumb sie nicht sollen zu hurerei gebraucht werden; denn das ist der geist, der von dem herren kompt, zu dem er auch wider gehet' (pp. 191–2)
57. Recall that he also considered the possibility that God was present in the womb in his commentary on the Psalms (See Chapter 2).
58. *Paracelsus Sämtliche Werke I*, vol. 9, p. 212. Crowther wrote of the language of agriculture appearing in works on reproduction. She notes that Horst regarded the process of generation in the womb as mirroring the process of generation in the larger world ('"Go Forth and Multiply"', p. 919).
59. He suggests that the woman's health reveals the condition of the womb. Therefore, he surmises that a connection exists between woman and womb that was unclear at the beginning of *Opus Paramirum* and in some of the earlier works considered in this book. Because he suggests that the woman can alter the womb, he implies that the presence

of the Holy Spirit has no bearing on the health of the uterus. Though he wants to consider the womb a divine realm, because he insists on some divine tie to conception, his evaluation of the effect of the female body on the womb suggests that he aligns only the *samen* that comes from the male body with God. Furthermore, Paracelsus considers the influence of the mind on the womb as analogous to the relationship of body and womb. He writes that a woman's imagination can benefit or harm the matrix, 'dan das der man den baum pflanzet und ferbet mit seinen henden, das tut die frau mit irer imagination, wie dan an selbigen enden und orten weiter gesagt wird. damit ist zuwissen das sich die matrix gesunt und ungesunt macht und helt, wie sie dan von dem darin sie ligt erhalten wird; nach demselbigen richt sie sich' (p. 212). The connection suggested between womb and imagination contradicts what Paracelsus wrote in *Buch von der Gebärung*, just as his comparison to arboreal growth differs from what had been said before, namely that human reproduction differs from plant reproduction. It becomes clear that Paracelsus's understanding of reproduction in the *Opus Paramirum* departs from his earlier works. Although as in his early works, he uses the term *samen* to refer to the material of conception, in the *Opus Paramirum,* only the man has *samen*. In this constellation, the woman's sole responsibility in reproduction is to carry the child in her womb. While Paracelsus regards the womb as having special ties to God (because of its development that is related to Creation and the presence of the Holy Spirit), he also suggests that the cause of undesirable births originate in the mother (being caused by her health or imagination). Thus, we must consider that Paracelsus's understanding of health is integral to his theory of human reproduction.

60. Though he still considers that man and women are born the same way, he emphasizes the difference of their bodies ' ...dan sie scheiden sich in den generationibus nicht von einander. darauf folget aber ein ander in der welt, ein ander im man, ein ander in der frauen. Darumb so folgt aus dem, dieweil die forme gibt ein ander welt zusein, so gibts auch das corpus' (p. 179).
61. 'So uns nu die aguen das ampt der cörper gibt, ein ander philosophiam der medizin für zunemen gegen der frauen, ire krankheiten nit zuvergleichen des mannes krankheiten und sich doch auf gleichnus der zeichen nit zu verlassen' (181) and 'darumb ist sie gespalten, das zwo welt da sein im menschen, der frauen und der man. Darumb das sie nicht gleich in den krankheiten, darumb so ist die arznei gescheiden' (p. 182).
62. *Paracelsus Sämtliche Werke I*, vol. 9, p. 82.
63. See especially pp. 186–7.
64. 'es befint sich kein krankheit im man, vor dem und ehe die frau beschaffen ist worden' and 'daraus dan zuwissen ist, dieweil alle krankheiten erst nachfolgend komen seind, das aus den frauen alle unser krankheiten herkommen und unser gebresten, elend und jamer' (pp. 188–9).
65. This tract can be found in *Paracelsus Sämtliche Werke I*, vol. 10, pp. 648–59. Unable to date this text definitively, Sudhoff proposed the year 1536. The introduction mentions 'Geben zu München...'. Sudhoff assumes that Paracelsus was in Munich at the end of 1536. Furthermore, he prefers 1536 because of Paracelsus's stated interest in astrology, which is also present in one of his most famous tracts, the *Astronomia Magna* (1537). See: *Paracelsus Sämtliche Werke I*, vol. 10, p. xlvi.
66. 'und am lezten do alle ding beschaffen seind gewesen und nichts mer not ist gewesen, do hat er den menschen beschaffen. non seind zwei do zu merken in der schöpfung, erstlich das er alle ding aus nichten beschaffen hat alein durch das wort, als alein den menschen

nit, den hat er aus etwas gemacht, das it er hat in gemacht aus iner masssa die ein corpus gewesen ist, ein substanz und etwas' (*Paracelsus Sämtliche Werke I*, vol. 10, p. 648.).
67. *Paracelsus Sämtliche Werke I*, vol. 10, p. 648.
68. *Grimms Wörterbuch Band*, vol. 12, pp. 1708–15.
69. ...geschaffen aus der welt, nicht nach ir gebildent, sonder nach gottes bilnus geformirt. Aber in im seind alle der welt eigenschaften beleiben, und die hat der Mensch in im (p. 649).
70. *Paracelsus Sämtliche Werke I*, vol. 10, p. 650
71. Ibid., p. 650

5 Paracelsus's Theory of Embodiment in the Popular Press

1. It is important to keep in mind that only a very small portion of Paracelsus's writing was published during his lifetime. About ten years after his death in 1541, word of his successful cures had spread throughout Europe and physicians began searching for the manuscripts he left behind as he moved throughout German-speaking regions. Today, scholars recognize the second half of the sixteenth century as the period of Paracelsus's greatest influence in the field of medicine and, particularly, in chemical medicine. See: A. Debus, 'Paracelsianism and Chemical Philosophy' in O. P. Grell (ed.), *Paracelsus: The Man and His Reputation His Ideas and Their Transformation* (Leiden, Boston, MA and Köln: Brill, 1998), pp. 225–44 and Debus, *The Chemical Philosophy*.
2. To this end, I draw on previous scholarship that has superbly explained the role of the printing press and the recognition of print as a business opportunity in the early sixteenth century. The work of Elizabeth Eisenstein, *The Printing Revolution in Early Modern Europe* (London: Cambridge, 1993) is indispensible and the work of William Eamon, *Science and the Secrets of Nature* (Princeton, NJ: Princeton University Press, 1994) provides background of how the occult sciences developed in Medieval Europe and how the printing press provided a new way to bring the secrets of the occult to the common man. Eamon does not consider Thurneisser's contribution to this field, although this chapter will demonstrate that Thurneisser did, indeed, play an important role.
3. My approach agrees with Jole Schakelford's interpretation of Paracelsus, however I focus on more popular vernacular German interpretations in the latter half of the sixteenth century. In a *Philosophical Path for Paracelsian Medicine* (Copenhagen: Museum Tusculanum Press, 2004), Schakelford explains how the Danish physician Petrus Severinus (*c.* 1540–1602) played an important role in spreading Paracelsus's philosophy of medicine throughout Europe. Bruce Moran also briefly outlines the work of some of Paracelsus's followers, though not including those I explore here, in *Distilling Knowledge*, pp. 81–98.
4. Thurnheisser and Hellwig wrote many popular books related to healthcare and understanding the body. Yet, even though Thurneisser and Hellwig are recognized paracelsian physicians few scholars have explored their impact on medical culture in their respective time periods, yet to do so would clarify the significant reach of Paracelsus's ideas. Laqueur's famous study on the cultural creation of sex differentiation, *Making Sex: Body and Gender from the Greeks to Freud,* argued that Paracelsus's work was so unique and singular that it had no bearing on how people thought about the body. Laqueur's underestimation of the importance of Paracelsus's medical ideas in the early modern period may stem from the lack of research on Paracelsus's followers that was available to him at the time he was researching his book. The available Paracelsus scholarship in print at the

time, demonstrates this unfortunate omission. For example, Allen Debus in *Chemical Philosophy* devoted scarcely a full page to Thurneisser's distillation process and, because of the scope of his project (sixteenth and seventeenth centuries), Debus did not consider Hellwig. Indeed, Debus directed our attention toward English and French Paracelsians. Pagel included a short section on German-speaking Paracelsians in *Paracelsus*, pp. 78. Pagel briefly focused on Thurneisser's thoughts on urine testing writing, 'Thurneisser's descriptions of the distillation of urine largely paraphrase the theory of catarrh in chemical terms. There are vapours ascending from the abdomen to the brain, which are condensed in the top part of the alembic – human chemical – and flow down to organs having ducts that are liable to be obstructed by the condensed and coagulated material (tartar)'. Pagel certainly recognized Thurneisser's success. Yet, while Thurneisser's contribution to the field of urinalysis is significant to the field of chemical medicine, the reader of Pagel's text does not get a sense of the widespread circulation of Paracelsus's ideas fostered by Thurneisser. This may explain why earlier ideas about a chemically based two-sex model of the body were overlooked. When Laqueur looked for secondary sources on Paracelsus to get a sense of his philosophy of the body, it is likely that he only found reference to the remarkable pharmaceutical and technological expertise that grew out of Paracelsus's approach to medicine. Indeed, the Paracelsus scholarship available at the time Laqueur published his work focused only on Paracelsus's influence outside German-speaking regions: England, France and Denmark. Shortly after Laqueur's book was published, research on Paracelsus's followers emerged as articles within larger Paracelsus studies and these articles tended to focus on the influence of Paracelsus's chemical philosophy. Only recently, has the writing of Paracelsus's followers gained greater attention through the work of Wilhelm Kühlmann and Joachim Telle (eds), *Der Frühparacelsismus*. Part 1 (Tübingen: Max Niemeyer, 2001) *Der Frühparacelsismus*. Part 2 (Tübingen: Max Niemeyer, 2004). Kühlmann and Telle edited two large volumes of letters written by some of Paracelsus's most famous sixteenth-century followers, like Toxites and Adam von Bodenstein, among others. The quote found at the beginning of this chapter is an excerpt from one of the letters they included in their two-volume collection. The many letters and the observation by Georg Joachim Rheticus (quoted at the beginning of the chapter) demonstrate the recognition of Paracelsus's influence in the second half of the sixteenth century. However, while Kühlmann and Telle's work provides us with resources for studying the letters of sixteenth-century Paracelsians, the task of framing these narratives within an understanding of Paracelsus's legacy remains.

5. Excerpt of the letter exchange with Toxites appears in Kühlmann and Telle (eds), *Der Frühparacelsismus*, vol. 2 (Tübingen: Max Niemeyer, 2004), pp. 437–8. Thurneisser also mentions reading Paracelsus's works during his apprenticeship in his book-length response to criticisms of his divorce. The work was reprinted by Will Erich Peuckert under the title *Der Alchymist und Sein Weib* (Stuttgart: Frommans, 1956). The correspondence in the 1570s with the famous paracelsian, Toxites (1515–1581), reveals Thurneisser's familiarity with paracelsian chemical practices. It is also a clear indication that Thurneisser read some of Paracelsus's works. In a letter dated 20. September 1576 sent Thurneisser a text from Paracelsus (Toxites calls him by his first name, Theophrastus) on plague. In the same letter, he describes his own experience with a distillation based on paracelsian methods. Toxites implores Thurneisser to let him know how the process works for him. This exchange of information affords us a glimpse of how paracelsian chemists/physicians tested what they learned from Paracelsus's text. In this case, the distillation that they call 'Spagiria' following the terminology used by Paracelsus in

his *Opus Paramirum*, is tested by Toxites in Frankfurt and by Thurneisser in Berlin. The 1576 letters from Toxites to Thunheisser also reveals that they exchanged Paracelsus's texts. Toxites is particularly concerned with retrieving a text called *Herbarium* from Thurneisser.

6. The definition of the term paracelsian to describe physicians generally refers to those who subscribe to Paracelsus's approach to the human as a microcosm and those who rely on chemical medicine. However, there has been some dispute about the exact meaning of the term. The association of paracelsians to chemical medicine was strengthened in the mid-twentieth century due to research by Walter Pagel, Allen Debus and earlier by Robert Multhauf in 'Medical Chemistry and "the Paracelsians"' *Bulletin of the History of Medicine*, 28:2 (1954), pp. 101–26. Peter Severinus (1542–1602) is perhaps the best known paracelsian physician of the time. His work *Idea medicinae philosophicae* appeared in 1571. See: O. P. Grell, 'The Acceptable Face of Paracelsianism: The Legacy of Idea Medicinae and the Introduction of Paracelsianism into Early Modern Denmark' in O. P. Grell (ed.), *Paracelsus: The Man and His Reputation His Ideas and Their Transformation* (Leiden, Boston, Köln: Brill, 1998), pp. 245–67. Debus argues that travel was an important aspect of the paracelsian physicians' identity in *The Chemical Philosophy*, p. 228. Andrew Weeks questions this aspect of paracelsian physicians' identity in his introduction to *Paracelsus: Theophrastus Bombastus von Hohenheim 1493–1541: Essential Theoretical Writings*. (Leiden: Brill, 2008).
7. For more information on Bodenstein's work see: Grell , 'The Acceptable Faces of Paracelsianism'. Frank Anderson suggested that Huber read Paracelsus's works as J.R. Partington had previously suggested in A History of Chemistry Vol.2 See: Frank Anderson, *An Illustrated History of the Herbals* (New York: Columbia University, 1977), p. 181 and J. R. Partington, *A History of Chemistry Vol. 2* (London: Macmillan, 1961), p. 152.
8. Gabriele Spitzer brought Thurnheisser's business prowess to our attention in her book *Und der Spree führt Gold. Leonhard Thurnhesser zum Thurn Astrologo-Alchimist-Arzt und Drucker im Berlin des 16. Jahrhunderts*. (Berlin: Staatsbibliothek zu Berlin Preussischer Kulturbesitz 1996), p. 26
9. Sometime after his expulsion, around 1548, Thurneisser traveled to England and then went to France. Later, he returned to Germany and fought under Count Albrecht von Brandenburg-Kulmbach until he was wounded on July 9, 1553 near Hannover and spent time in prison. Between 1555 and 1558 he worked as a goldsmith in Strassburg and other areas. By some accounts, during this period he returned to his hometown, Basel, in 1555 and in either 1556 or 1557 he married Margerete Müller (perhaps a widow). He lived with her a short time and then had an affair with Lorenz Viy. Through this affair he had a son, Hans Jacob (born 1559). In 1558 he purchased a coal mine in Tirol and also earned a living by giving advice on minerals to wealthy citizens of the area. For various reasons, Thurneisser was forced to travel throughout central Europe; however it seems that his travels were exaggerated to include the Far East and Africa. Some of this biography I have pieced together from his book *Confirmatio Concertationis*, some from A. Hirsch's *Biographische Lexikon der hervorragenden Ärzte aller Zeiten und Völker* (Berlin-Wien: Urban and Schwarzenberg, 1934), vol. 5, pp. 582–3, and some from the work of Gabriel Spitzer, author of *...und die Spree führt Gold* pp. 11–23.
10. W.-E. Peuckert (ed.), *Der Alchemist und Sein Weib: Gauner und Ehescheidungsprozesse des Alchymisten Thurneisser* (Stuttgart: Günter Holzboog, 1956).
11. Pagel cites this famous letter written by Oporinus in *Paracelsus*, pp. 29–30.

12. Many print images and some stained glass portraits remain. See: P. Boerlin, *Leonhard Thurneisser als Auftraggeber: Kunst im Dienste der Selbstdarstellung zwischen Humanismus und Barock*. (Basel and Stuttgart: Birkhäuser, 1976).
13. T. Nummedal. *Alchemy and Authority in the Holy Roman Empire* (Chicago, IL: University of Chicago Press, 2007), pp. 40–72. She explains another instance in which Thurnheisser attempts to rectify the assumed deceitfulness of alchemists in his *Magna Alchimia* (1538). See p. 63.
14. P. Boerlin, *Leonhard Thurneisser als Auftraggeber* (Basel: Birkhäuser, 1976), p. 16
15. In Hirsch's encyclopaedia entry he maintains that Thurneisser's interest in paracelsian medicine served purely to profit from specific secrets of medicine found in Paracelsus's texts, '*Thurneisser gehört zu der jenigen schlechten Sorte der Paracelsisten, denen er nur darum zu tun war, ihres Lehrers geheime Künste zum Zweck eigener Bereicherung zu erlernen und auszunutzen*' 'Thurnheisser belongs to the worst kind of Paracelsians, one who just used the secret art for the purpose of becoming rich' (vol. 5, p. 583). This assessment of Thurneisser as a follower of Paracelsus echoes the sentiments expressed by Walter Pagel: 'Leonhard Thurneisser zum Thurn, der bekannte Abenteurer vom reinsten oder vielmehr unreinsten Wasser, ein Paracelsist jener schlechteren Sorte, die ihrem Meister keine Ehre bereitete, Leibarzt Johann Georg's seit 1578, bis er in Folge seines schandalösen Treiben 1582 in Ungnade fiel, ein Mann, der durch die unsaubersten Geschäfte, durch Verkauf von Geheimmitteln, Fabrikation der medicinische-astrologischen Kalender, die von Blödsinn strotzen, zu Verbreitung des medicinisch-astrologischen Kalender, die von Blödsinn strotzten, zur Verbreitung des medicinischen Aberglaubens unter der Menge überdies durch sein eigenes, schlechtes Beispiel zu moralischen Deparavation seiner Umgebung zm moisten beigetragen hat', J. L. Pagel, *Entwicklung der Medicine in Berlin* (Wiesbaden: Bergman, 1897), p. 8.
16. P. Boerlin, *Thurneisser als Auftraggeber* (Basel and Stuttgart: Birkhäuser, 1976.), pp. 16–17.
17. G. Spitzer, *und die Spree führt Gold*: *Leonhard Thurneisser zum Thurn Astrologo-Alchimist-Arzt und Drucker im Berlin des 16. Jahrhunderts* (Berlin: Staatsbibliothek zu Berlin Preussischer Kulturbesitz, 1996), pp. 70 and 84.
18. Leonhard Thurnheisser zum Thurn. *Concertatio Concertationis*. MS St Louis, Becker Medical Library Washington University. This version has leather binding and some hand colouring (green). Title page, confirmation concertationis
19. G. Spitzer, *...und die Spree führt Gold. Leonhard Thurnhesser zum Thurn Astrologo-Alchimist-Arzt und Drucker im Berlin des 16. Jahrhunderts*. (Berlin: Staatsbibliothek zu Berlin Preussischer Kulturbesitz 1996).
20. K. R. Encarnación, 'The Proper Uses of Desire: Sex and Procreation in Reformation Anatomical Sheets', in A. L. McClanan and K. R. Encarnación (eds), *The Material Culture of Sex, Procreation, and Marriage in Premodern Europe* (New York: Palgrave, 2002), p. 223.
21. For this information I am indebted to the discussions held at the Women Health and Healing in Early Modern Europe Conference organized by Mary Fissell and held at Johns Hopkins University April 21, 2006. In particular I would like to thank Kathleen Crowther for her dedicated research on the topic. In addition, G. Spitzer in her book on Thurnheisser suggested that Thurneisser's inclusion of flap anatomies may have been the first of its kind and she correctly points to another paracelsian follower's inclusion of flap anatomies after Thurneisser's publication: Bartisch. For more information about anatomical renderings of the sixteenth and early seventeenth century see: A. Carlino

Paper Bodies: A Catalogue of Anatomical Fugitive Sheets 1538–1687, trans. N. Arikha (London: Wellcome Institute for the History of Medicine, 1999), especially pp. 108–9. Carlino argues that the anatomical illustrations served two purposes: to remind educated physicians what they saw when they witnessed a dissection and for uneducated for barbers to display in their shops as evidence of their knowledge of bloodletting. Additional information on lift-a-flap prints including religious prints of the reformation era see: Suzanne Kathleen Karr Schmidt, 'Art – A User's Guide: Interactive and Sculptural Printmaking in the Renaissance' (PhD dissertation, Yale University, 2006).

22. *Concertationis*, p. 39r. (male) and p. 40v. (female). The reference to the difference in women's breasts, which is exaggerated in the visual depiction, has a marginal gloss that reads, 'Weiber grosser brüste dann die Menner' p.40r. I assume that this marginal gloss, like the line drawing with the enlarged breasts and nipples, are meant to entice the presumably male reader to examine the page.
23. W. Eamon, *Science and the Secrets of Nature: Books of Secrets in Medieval and Early Modern Culture* . (Princeton, NJ: Princeton University Press, 1994), p. 102.
24. To depict the change in anatomical representation of the eighteenth century, Laqueur drew on the work of L. Schiebinger, 'Skeletons in the Closet: The First Illustrations of the FemaleSkeleton in Eighteenth-Century Anatomy'. *Feminism and the Body*, ed. Londa Shiebinger (Oxford: Oxford University Press, 2000), pp. 25–52. Schiebinger claimed that the drawings of female skeletons first appeared between 1730 and 1760 in England, France, and Germany. According to Schiebinger, women's inferior status was denoted in anatomical drawings by the way the head was drawn as relatively small in proportion to her body. She reads the signs of a distinctly female body in the emphasis on the width of female hips, which she argued began in the eighteenth century. Michael Stolberg refuted Schiebinger's claim. 'Critiques and Conventions'. In this article, Stolberg shows that the female skeleton and explicit sexual dimorphism appeared in the sixteenth century, namely in 1583 in Felix Platter's *De corpus humani structura et usu libri III*. Stolberg notes that Platter's work was remarkable in that it labelled the anatomical representation of the female body to show how it differed from the male body. Stolberg's work focuses on an academic text, written in Latin. Thurneisser's book, published before Platter's in 1576, demonstrates that a two-sex model was accessible to a wide range of readers.
25. Even in Vesalius's famous *De Humani Corporis Fabrica* of 1543, a Latin text used by medical students, ambiguity about the connection between the spermatic ducts and the testes is present. There is no visual representation of the connection between testicles and penis.
26. Du Laurens, Andre Histori Anatomica Humani Corporis (Frankfurt: Becker, 1600), p. 169. Becker Medical Library, Washington University. The text was first published in Erfurt in 1595.
27. C. Estienne. *La dissection des parties du corps humain* (Paris: Simon de Colines, 1546), p. 202. Becker Medical Library, Washington University.
28. The anatomical depictions of male and female represented as Adam and Eve was common in the early modern period. See: K. Crowther, 'Creating Adam and Eve: Body, Soul and Gender in Sixteenth Century Germany' (PhD dissertation, Johns Hopkins University, Baltimore, MD 2001).
29. In the male distillation chart it reads, 'Der mitlen sitrnader gelegenheit und zerteilung unde und oben eine X gleich', it is the same on the female distillation chart. The next 'Zerteilunge der schaff arterien ist biss ins haupt durch veil ast und zwig geteilt', again this is the same in both, but on the female chart one finds an explanation: 'welcher die

ader durch welche die Menstrualische feuchte in der mutter hals geleit wird/wie auch die schamander und narungsader'.

30. He regards the womb as wondrous and writes with awe about its importance and the connection to God. Again his description of the womb and his understanding of how conception occurs echoes the understanding put forth by Paracelsus earlier in the century. He differs only in his attempt to ponder the special little woman's part, the vagina.
31. 'Zusammen kunst der mancherley aderlein so vom ganzen Cörper sich versamlen/ und zu der Bermutter des weibs gand/ do die sich widerum unmancherley sprosslein aufftheilen/ durch welche mit hilff der narunge so durch anderen geng von den brusten darzu gand/ die Frucht ernert wird'. The kidneys magically collect material from the entire body and the pregnant mother will have milk in her breasts that serve as the fruit of nourishment. L. Thurnheisser, *Confirmatio Concertationis*. (Im Grauen Kloster, 1576). Becker Medical Library Rare Books Collection, Washington University St. Louis, Missouri, U.S.A. (hereafter *Confirmatio Concerationis*), p. 32.
32. *Confirmatio Concerationis*, p. 34r.
33. Ibid., p. 35r.
34. He wrote, 'The old philosophers did not even know a name to create for this part'. 'Alte philosophi haben diesem glidt keinen Eigentlichen Nahmen gewust zuschopfen' (page 36r). On the same page one finds four diagrams that show 1. The proximity to the bladder (this image uses lift-a-flap). 2. A gravid uterus with foetus facing forward. 3. A gravid uterus with foetus facing the spine 4. a lift-a-flap uterus showing the proximity to a gravid uterus. All diagrams are labelled with letters that correspond to the distillation chart.
35. *Confirmatio Concertationis*, p. 34r and 35v.
36. The passage reads: 'doher alwegen volget das des Weibs Geist schwecher darumb muessen inn allen arbeitten der proben durchauss die harn ob sie Weiblich oder Mennlich/ arzu ob sie Alt/ Mittel/ oder Jung seien underscheiden warden. Das aber das weib inn empfindunge Leblicher Geister schwecher dargegen inn wircklicher Krafft derselbigen vermöglicher sey/ wird an dem gespierdt/ das die allein durch ire Imagination oder einbildunge/ ethwas das schön erschaffen und geformiret ist/ inn einem augenblick'. Therefore it follows that the woman's spirit (or intellect) is weaker and therefore one must work to test if the excrement (urine or stool) is feminine or masculine, medicine should be adjusted for old, middle age, or young. Because women's constitution is weaker and because she is so influenced by her imagination she should be shown something pretty during this moment. *Confirmatio Concertationis* p. 35r.
37. Thurnheisser's biblical account of human birth echoes the sentiments expressed by Paracelsus in his book about birth, but expands the idea to include the liver as a significant contributor to the materials for gestation.
38. This differentiation was brought to my attention by my husband, Geoffrey Cislo, M.D. assistant professor of Medicine at Washington University in St Louis. Thanks Geoff!
39. 'von Schöpfung und Vereinbarung des Menschlichen Cörpers/ Facultet des Arztes/ von Uhrsprung der Kranckheitten/ und ersten Natürlichen Anfengen aller deren Dingen/ Leonart Thurneisser zum Thurn das Erste Buch' 'About the Creation and Union of human bodies/ The Work of Doctors/ About the Origins of Illness/ and the first natural beginning of things/ by Leonart Thruneisser zum Thurn, the first book. *Confirmatio Concertationis* p.1r.
40. 'So befindt sich doch/ das nicht allein die gestalten/ aller dieser Creaturen/ einziger weiss/ sunder auch derselbigen zertheitle/und besondere glieder/ sambt iren gewerblein/ von diesem Dreynahmenden (und doch Enzigen himelischen und aller gewaltigsten)

werkmeister/ inn dem Gott gleich geformbten Cörper des Menschen/ also gar eigentlcih fürgebildet/ und mit socher Proprietet/ abgemahlt und entworffen seindt/ das nichts inn diesem grossen Globo/ und umbkreis der ganzen welt/ es sey im himel oder auff Erdtrich/ gefunden wierdt/ das nicht auch inn diesem bildt/ umb deessen willen es dann alles erschaffen/ welcher auch ein Subiectum Philosophae, Astronomie udn Elementorum ist/ Klarlich anzeigt/ und sichtbarlcih zuerkennen/ oder etwan an einem siner Membris und theilen zufinden sey. *Confirmatio Concertationis* p.2r.

41. This gloss can be found on page 2v, in the section titled, 'On the beginning of Things' 'Von den anfengen der dingen'.
42. The theory is presented as a chart: Aus einem (Essentialischen oder himlischen, Geistlichen oder Elementischen, Naturlich oder Leyblichen} Cörper Erbaut/ und Erschaffen 'Out of one: essential or heavenly, spiritual or elemental, natural or bodily'. *Confirmatio Concertationis* p. 2v. The three qualities in brackets are listed vertically, so that the entire schema is indented in the text and thus visually captivating.
43. See Chapter 4.
44. He writes, 'Die Gute, Verkerte Bose ist: Ewig Lablich, Ewig verdamlich/oder verderblich/ Ewig Todlich' *Confirmatio Concertationis* p.3r.
45. *Confirmatio Concertationis*, p. 3r.
46. Ibid., p. 3v.
47. Marketing medical knowledge was not new to this period. There is evidence to suggest that the tradition began in the fifteenth century when publishers recognized the marketability of health. See: Isidro J. Rivera, *Negotiation of Scientific Discourse in the First Printed Edition of the Historia de la donzella Teodor* (Toledo: Pedro Hagenbach, ca. 1500) *Hispanic Review*, Vol. 66, No. 4 (Autumn, 1998), pp. 415–32.
48. The basilisk was a well-known wonder of uncertain origins in Europe as early as the thirteenth century. Albertus Magnus (1193–1280) rejected the long-held belief that a basilisk was born from a cock's egg in *De animalibus* . See Laurence Breiner, 'The Basilisk', in Malcolm South (ed.) Mythical and Fabulous Creatures (Wesport, CT: Greenwood Press, 1986), pp. 113–22. Fascination with the Basilisk must have continued for some time because Paracelsus contributed his own hypothesis about its origins in *Liber de Homunculus*, 'dan on den dotter aus dem geborn wird der basilisk, aus dem weißen das weiblin eins basilisken' (*Paracelsus Sämtliche Werke I*, vol. 14, p. 327) and after Paracelsus's death Jacques Grévins (c.1540–1570) also discussed its origins in *Buch über die Gifte*. See: Thomas Cramer, 'Der Umgang mit dem Wunderbaren in der Natur' *Knowledge, Science and Literature in Early Modern Germany* eds. Gerhild Scholz Williams and Stephan K. Schindler (Chapel Hill, NC: University of North Carolina Press, 1996), pp. 164–5. Cramer explains how interpretations of the origins of the basilisk from the medieval to the seventeenth century reveal broader assumptions about nature.
49. *Paracelsus Sämtliche Werke I*, vol. 3, pp. 247–92.
50. Karl Sudhoff, a twentieth-century compiler of Paracelsus's works argued that there was reason to suspect that this text may not be by Paracelsus.
51. Karl Sudhoff explains the history of how these texts were first made public by Adam Bodenstein in Basel in the 1570s. He cites the examples of the distribution noted by Huser in his quarto edition. See Sudhoff, *Theophrast von Hohenheim gen. Paracelsus Sämtliche Werke I*, vol. 3 (München: R. Oldenbourg, 1930), pp. xii–xiii and particularly the manuscripts *De Longa Vita* on pp. xxxii–xxxiv.
52. Dann was ist wunderbarlichers dan dieses/ dass aus einem wol (und von der dritten und letzten tewunge) weiss gekochten geblüet/ und feüchte/ welche auss den Edelsten

und fürnembsten theilen dess ganzen Leibes versamlet/ so es in diss gliedt durch Menschliche ubunge springt/ darm so bald beschlossen/ und von stundt an Innerthalb siben tagen/ inn solche treffentliche underscheidung dreyer ungleicher gliederen/ Nemlich eines hirns/ einer Leber/ und eines herzens verendert/ und nach der Separation ein bildtnus irer eignen form gleichmessig erwachsset/ auch über 46. Tag das Leben/ inn 90. Tagen aber die bewegung empfahen thut/ Ja ob nun gleich dieses dass die frucht mit so vil underscheidner gliedern/ Inn und ausserthalb des Leibes/ als Bein / Haut/ Haar/ Fleisch/ Bludt und feüchte bezieret wirdet/ wunderbarlich ist/ So ist doch dieses noch vil wunderbarlicher/ und Menschlicher vernunfft nicht wol zufassen/ das der Mensch der sich also susamen geschumcket/ un mit dem angesicht gegen Mütter Rücken wendet/. (*Confirmatio Concertationis*, p. 36R)

53. Kathleen Crowther called this a metaphor of craft work in her analysis of other sixteenth –century vernacular texts. See: Kathleen Crowther-Heyck. '"Be Fruitful and Multiply": Genesis and Generation in Reformation Germany. *Renaissance Quarterly*, 55 (2002), pp. 904–35.
54. The poem reads:

    Ihr meister weis der Medicin/
    Geben anzeigung mir/ und schin/
    Damit ich werdt einer sach bericht/
    Von der ich vor nie hab gedicht.
    Dass ich erfind ursach/ und grundt/
    Einer frage/ so hernacher kundt/
    Von edlem geschlect/ und schönem Leib/
    Des Vesten Ludwig Sturmen weib/
    Frau ann von Enndingen gebnorn/
    Hatt jetzt zu Strasburg bey acht Jarn/
    Gelitten Kranckheit manchereley/
    Ein solch gethon/ und grausam gschrey
    Gar offt inn irem Leib gehan/
    Wol zwanzig schritt von ir hindan/
    Hort man das krachen/ und den ton/
    Sie must offt heim von Kirchen gon/
    Verlassen auch das heylig ampt/
    Das sie sich vor de Leutten schambt/
    Darmit hatt sie wehe/ schmerzen gross/
    Die gute fraw wenig verdross/
    All fleiss/ Ernst/ kosten zlegen an/
    Ob sie doch jemands hülff mocht han/
    Nach Gelerten Arzten hat sie gestalt/
    Inn zween zufell ist sie gefelt/... From pages 51r and 52v.

55. Hirsh, *Biographisches Lexikon* 2te Auflage (Berlin 1931) vol. 3, pp. 149–50.
56. For a full listing of Hellwigs works see: S. Sander, '"Aufklärung vor der Aufklärung': Zum poplulärmedizinischen Werk des Arztes und Bestsellautors Christoph von Hellwig (1663–1721)', *Medizinhistorisches Journal*, 34 (1999), pp. 283–308.
57. He explains this philosophy in the introduction dated 12 September 1729 of his book published under the pseudonym, V. Kräutermann, *Die wohl underwiesene Apothecker*

oder Gründliche Anleitung zur heutigen Apothecker-Kunst (Arndstadt und Leipzig: Johann Jacob Beumelburg, 1735).

58. I stumbled upon this discrepancy in the dating while working on the pharmacological text from 1735. At first, I thought that Valentine Kräutermann was perhaps not Christoph von Hellwig and I considered omitting the text from this chapter; however an article by Sabine Sander explained that it was not uncommon to find Hellwig's introductions post-dated in the fourth and fifth reprinting of his books. See: S. Sander, '"Aufklärung vor der Aufklärung": um poplulärmedizinischen Werk des Arztes und Bestsellautors Christoph von Hellwig (1663–1721)' *Medizinhistorisches Journal*,34 (1999), p. 250
59. V. Kräutermann, *Curieuser und vernünfftiger Urin-Arzt: welcher eines Theils lehret und zeiget wie man aus dem Urin nicht allein die meisten und vornehmsten kranckheiten des menschlichen Leibes nach gewissen Kunstregeln der Medicin gründlich erkennen, sonder auch wie ein vernunfiges Judicum davon zu fallen sey*. (Ardstadt: Johann Jacob Beumelburg, 1748).
60. V. Kräutermann, *Curieuser und vernünfftiger Urin-Arzt, Welcher eines Theils lehret und zeiget, Wie man aus dem Urin nicht allein deie meisten und vornehmsten Krankheiten des menschlichen Leibs nach gewissen Kunstregeln der Medizin gründlich erkennen, sondern auch, wie ein vernünfftiges Judicium davon zu fallen sey*. 4 edition Arnstadt: Johan Jabob Beumelburg, 1748), p. 102
61. Kräutermann, *Curieuser und vernünfftiger Urin-Arzt*, p. 112.
62. Ibid., p. 113.
63. He writes, 'Wenn der Mann mit dem Weibe des Nachts der Liebe gepflogen, so ist des Weibes harn trübe, und des Mannes Saame liegt an dem Grunde' 'If the man and woman make love in the night, the woman's excrement will appear murky and the reason for this is the man's sperm'. *Curieuser und vernünfftiger Urin-Arzt*, p. 112
64. Hellwig also produced an anatomy text. The text we will consider here does not bear a date, but the dedication is dated 16 May 1720. The title reads: *Nosce te Ipsum, anatomicum Vivum, oder: Kurz gefastes doch richtig gestelltes anatomisches Werk, Worinnen die ganze Anatomie, nebst ihrer Entheilung deutlich zu finden. Jedwedem nützlich, anmuthig und erbaulich zu lessen, nebst nöthigen kupffern/ wovon die Invention ganz sonderlich/ indem man die viscera heraus nehmen/ nach denen Tabellen wohl betracten, sich aus dem Tractätlein darbey informiren, und wiederum einfezen kan, wovon die Vorrede und Information ein mehreres Licht geben. Nebst beschreibung tötlicher Wunden; mit angefügtem Register auf Begehren zum andernmahl zum Druck befördert.* Condensed title: Know thyself, the anatomy of life or: Short but correctly diagrammed anatomical work, wherein the whole anatomy and healing are explained. Useful for everyone, charming and educational reading, for curious minds that want to learn about the workings of the body using tables and tactile learning... (Frankfurt and Leipzig: H.P. Ritscheln, *c.* 1720) The introduction explains that Hellwig's patron was Herr D. Joh. Phillip Eysel, Philos. & Med. D. Anatom., Chirurg. &Botan, who introduced him to anatomy and guided him in preparing the book. It further explains that the title, Nosce te Ipsum (Know thy self) refers to the connection between theology and the body. The text invites the reader to learn about himself to enrich his spirituality by learning more about the beauty of God's creation. The narrative portion of the short book includes a twelve page description of how to use the text, followed by three pages of lift-a-flap anatomies, tables that correspond to the letters on the anatomy charts and finally, an index. The text contains lift-a-flap anatomy, though less sophisticated than those found in Thurneisser's and they appear to have been influenced by Remmelin, or they may simply be a crude copy of

Remmelin's very popular anatomy book. Remmelin of Augsburg along with Michael Spahere of Tyrol produced an anatomy text that compared the male and female bodies. The work was called, *A Survey of the Microcosm or, The Anatomy of the Bodies of Man and Woman*. It was first published in 1670, though it was reprinted at least five times with the last printing in 1738. In other words, the Remmelin text circulated at the same time as Hellwig's anatomy. In both texts, the male and female bodies are situated on Grecian-looking pedestals on the same page. The male figure gazes toward the female and we see his front side, the female figure gazes toward the male and we see a dorsal view. Aside from the long hair and the delicate facial features, the female anatomical model looks just like the male. Both have wide shoulders, and the width of the hips of the female model appears the same as the male. A separate lift-a-flap abdomen appears to describe the female reproductive parts, but even here the hips and pelvis are narrow like a man's. Under the first flap one finds a rough drawing of what should be vulva, labia and clitoris, instead one finds a rather ambiguous darkly shaded area with the labels, e,f, and g, which corresponds to the table where one reads: 'Der Eingang, Die zweene Berge, Colles, Ein zart Hautlein Hymen gennant'. So, we have the entrance, the two hills and a hymen. Clearly, the intended audience is male, since the vagina is described as an entrance. Another full anatomical figure of a woman has a cherub face and a male body. Overall, Hellwig's descriptions and visual representations of the body are less sophisticated than those in Thurneisser's text. Recalling that Thurneisser's text circulated nearly a century before Hellwig's one may ask what happened? It appears that Hellwig was trying to address and, maybe not offend, an audience educated in the field of medicine, so he used Galenic style imagery that would appeal to those familiar with classical texts and the kind of anatomical illustration used in medical schools. Hellwig was probably introduced to such drawings during his own time in medical school. The title suggests an approach to the body that would agree with Paracelsus's philosophy of the body as a microcosm, but the text itself does not dwell on the meaning of the body.

65. 'Das erste Tagewerk des Allmächtigen Gottes, bey Erschaffung der Welt, war das Licht, damit derselben Schönheit in einem dunckeln Abysso nicht verstecket bleiben möge. Das Auge ist das Licht des Microcosmi, und ein Zeuge, dass das, was Gott gemacht, alles gut sey. Durch das Auge betrachten wir die grossen Wunder Gottes, des Firmaments, der Welt, deren Ordnung und Annehmlichkeit'. Quoted from: *Der Sichere augen und Zahn Arzt Oder accurate Beschreibung Aller und Ieden Augen und Zahn=Gebrechen, Nebst Deutlichen Unterrichte, soclen bey zeingen vorzukommen, oder auch glücklich zu curiren sind. Nebst nöthigem Register ans Licht gestellt von Valent. Kräutermann, Med. Pr.* (Arnstadt u. Leipzig: Johann Jacob Beumelburg 1737).
66. Images of this text can be found at the following url: http://www.mclibrary.duke.edu/hom/exhibits/bartisch/ [accessed 10 January 2009].
67. My future research will explore in greater detail how the transmission of Paracelsian ideas passed from Thurnheisser to Bartisch.
68. In *Der sichere Augen- und Zahn- Arzt* (pp. 19–22) Hellwig goes on in great detail describing the way the eye works and all the aspects of the eye that one should recognize, without drawings. Next, he explains ways that one could damage the eye: wounds, bee stings, wasp stings. He explains that these occurrences could cause infection and then provides a recipe for a salve to remedy the problems. He follows the same pattern of description of the problem followed by causes when he addresses ailments of the eyelid. In Hellwig's books one often finds this format: explanation of the possible problems and salves to cure or lessen the illness. He includes remedies for cancer, glaucoma, cataracts,

loss of eyelashes and all illnesses of the eye that he has encountered, like something he calls black star, a kind of blindness, different from night blindness (another issue he addresses). The book also contains dentistry in the same fashion as the description of eye problems: first a detailed description of the problem followed by a remedy. All references to this text are: V. Kräutermann, *Der Sichere Augen und Zahn Arzt* (Arnstadt and Leipzig, 1737). Rare Books Collection, Becker Medical Library, Washington University. For more on the popularity of Paracelsus's approach to natural remedies see: Cooper, *Inventing the Indigenous*, pp. 21–50.

69. Sanders, 'Aufklärung vor der Aufklärung', pp. 268–70
70. V. Kräutermann, *Der Thuringische Theophrastus Paracelsus* (Arnstadt and Leipizig: E.L. Niedt, 1730).
71. This brief analysis of Hellwig's promotion of Paracelsus's ideas will be expanded in my future research on paracelsianism in the popular press.
72. Kräutermann, *Paracelsus*, p. 132.
73. The book contains remedies attributed to Paracelsus for arousing one's sex partner, how to increase ejaculation, how to stimulate an erection and other advice on sexuality. Nowhere in Paracelsus's oeuvre does such advice exist.
74. For more information on remedies for the care of loved-ones, and how care took place in the home see: Alisha Rankin, 'Duchess, Heal Thyself: Elisabeth of Rochlitz and the Patient's Perspective in Early Modern Germany'. *Bulletin of the History of Medicine* 82.1 (2008), pp. 109–44. Rankin also explains the high level of expertise among women who cared for their loved ones in: 'Becoming an Expert Practitioner' ISIS 98.1 (2007), pp. 23–53.
75. The well known psychologyist, C. G. Jung recognized the psychological concepts found in early modern alchemical texts in his book, *Psychology and Alchemy* (New York: Pantheon, 1953). For more information on this topic see: H. Schott and R. Tölle, *Geschichte der Psychiatrie: Krankheitslehren Irrwege Behandlungsformen* (Munich: C.H. Beck, 2006) and H. C. Eric Midelfort's *A History of Madness.*

Appendix

1. H. Rudolph, '"Paracelsus" Laien Theologie in traditionsgeschichtlicher Sicht und in ihrer Zuordnung zu Reformation und Katholisher Reform', in P. Dilg and H. Rudolph (eds.), *Resultate und Disiderate für Paracelsus-Forschung* (Stuttgart: F.Steiner, 1993), p. 86 and 87 writes about how Buch von der Gebärung was one of the earliest tracts that reveal Paracelsus's theological interests and he compares the viewpoints expressed in Pars de statu humano to many of Paracelsus's theological tracts from the 1530s. I agree with this comparison and think that the tract, Pars de statu humono should be read separately from Buch von der Gebärung.
2. Grimm's dictionary defines the usage of 'empfindlich' during this time to mean sensibilis (Latin) or perceptible.
3. Sudhoff noted that this text may be a fragment. Huser, one of the first collectors and publishers of Paracelsus's works, used a manuscript he acquired from a Dr. Homelius as a primary source for inclusion in his first volume of Paracelsus' s works. Huser's version appeared in print in 1589. Sudhoff transcribed exactly what he found in the Huser Quarto edition of 1589. I compared the texts side-by-side an noticed no significant differences. Sudhoff estimates that *Buch von der Gebärung...* was written around the same time that Paracelsus composed *Von den fünf entien*, thus dating it around 1520. I used

the Sudhoff version, which appears in volume 1 of his collection. The corresponding page numbers are provided in brackets. In several instances I inserted stops to make the writing easier to read. Paracelsus often writes sentences that are thirty words long.

4. 'licht der nature' is an expression used frequently in Paracelsus's writing. He regards it as the building block of life.
5. This passage is confusing and fragmentary in the original. There it reads: Die andern, seine lerer und seine Schuler, haben sich eins solchen sinnes und klugheit dergleichen geflissen mit vilen umbfürendnen kunstreichen, sinnischen argumenten und irs gleichen herfürbringungen geübt zu gebrauchen, aber im kern der nature in staub und ein zerknitschter blum (*Paracelsus Sämtliche Werke*, vol. 1, p. 244).
6. The sentence continues: dergleichen herwiderumb verachten wir die sinliche listikeit, die sich philosophiam nent und ein geferbts gedünken ist, aber wol geblümt und ausküttet. Paracelsus uses a metaphor of the life of flowers to describe philosophy. Because the phrase is confusing in the original, I omitted it in this translation.
7. The following phrase was omitted due to its problematic phrasing in the original: aber wir wollens wie en vergebeni luft an uns streichen lassen und für rauschen, bis zertrent wird.
8. Paracelsus uses the word *ietlich* throughout his writing to describe beings that lack a special connection to God.
9. The sentence continues: 'die sonst die erkentnus one sie nit hetten, aber von wegen einer besonderen sel'. It is unclear but suggests that he's referring to beings who lack a soul.
10. This sentence is confusing in the original. It reads: *also solt ir merken, das im entpfintlichen gewechs ein ieglichs ding für sich selbs ein muter ist on ein man, und was menlich erkennet wird under inen, das ist alien auch ein muter, wan es tregt auch samen.*
11. Possibly a reference to the poet, Albius Tibullus (*c.* 55 BC– 19 BC) who wrote of his love affair with a woman named Delia.
12. In the original the word is crystalline. K. Sudhoff explains that the word means visible organs in *Paracelsus Sämtliche Werke I*, vol. 1, p. 409.
13. This passage is very confusing in the original: 'wan der same in der vergleichung kompt allemal auf ein ungleice wag, also das er nimmer instehet, das ein mensch möcht geboren warden, das seinem vater oder seiner muter gleich were wie sie dan seind'
14. He does use the word sperma here, but he seems to be talking about seeds.
15. Probably the *Volumen Paramirum*, another of Paracelsus's early works.
16. In this passage, he once again criticizes Aristotle's teachings, but the passage is too problematic to translate: die gute beischürer nennen die da Aristotlem und den Augustum einbilden, vermeinen sie wollen die selbigen wider machen'.

WORKS CITED

Primary Sources

Der Alchymist und Sein Weib: Gauner-und Ehescheidungsprozesse des Alchymisten Thurneysser. Ed. Will-Erich Peuckert (Stuttgart: Fr. Frommann, 1956).

Hellwig, Christoph von. (aka Valentine Kräutermann). *Curieuser und vernünfftiger Urin-Arzt: welcher eines Theils lehret und zeiget wie man aus dem Urin nicht allein die meisten und vornehmsten kranckheiten des menschlichen Leibes nach gewissen Kunstregeln der Medicin gründlich erkennen, sonder auch wie ein vernunfiges Judicum davon zu fallen sey*. (Ardstadt: Johann Jacob Beumelburg, 1748).

Hellwig, Chistoph von. *Nosce te Ipsum, anatomicum Vivum, oder: Kurz gefastes doch richtig gestelltes anatomisches Werk, Worinnen die ganze Anatomie, nebst ihrer Entheilung deutlich zu finden. Jedwedem nützlich, anmuthig und erbaulich zu lessen, nebst nöthigen kupffern/ wovon die Invention ganz sonderlich/ indem man die viscera heraus nehmen/ nach denen Tabellen wohl betracten, sich aus dem Tractätlein darbey informiren, und wiederum einfezen kan, wovon die Vorrede und Information ein mehreres Licht geben. Nebst beschreibung tötlicher Wunden; mit angefügtem Register auf Begehren zum andernmahl zum Druck befördert*. (Frankfurt and Leipzig: H. P. Ritscheln, *c.* 1720).

Hellwig, Christoph von. *Der Sichere augen und Zahn Arzt Oder accurate Beschreibung Aller und Ieden Augen und Zahn=Gebrechen, Nebst Deutlichen Unterrichte, soclen bey zeingen vorzukommen, oder auch glücklich zu curiren sind. Nebst nöthigem Register ans Licht gestellt von Valent. Kräutermann, Med. Pr.* (Arnstadt u. Leipzig: Johann Jacob Beumelburg 1737).

Kräutermann, Valentin (aka Christoph von Hellwig). *Der Thuringische Theophrastus Paracelsus* (Arnstadt and Leipizig: E.L. Niedt, 1730).

Paracelsus, 'De Invocatione Beatae Mariae Virginis' Katharina Biegger (ed) *'De Invocatione Beatae Mariae Virginis' : Paracelsus und die Marienverehrung* (Stuttgart: Steiner, 1990), pp. 150–84.

Paracelsus Sämtliche Werke. I. Abteilung. Medizinische, naturwissenschaftliche und philosophische Schriften, ed. K. Sudhoff, vols 6–9 (München/Berlin: O.W. Barth, 1922–5) vols 1–5 and 10–14 (Munich: R. Oldenbourg, 1928–33).

Paracelsus Sämtliche Werke II. Abteilung. Die theologischen und religionsphilosophischen Schriften, ed. K. Sudhoff and W. Mattheissen, vol. 1. (Munich: O.W. Barth, 1923).

Paracelsus Sämtliche Werke II. Abteilung Band III. Die theologischen und religionsphilosophischen Schriften, ed. K. Goldammer (Stuttgart: Steiner, 1986).

Thurnheisser, L., *Confirmatio Concertationis.* (Im Grauen Kloster, 1576). Becker Medical Library Rare Books Collection, Washington University St Louis, Missouri, USA.

Secondary Sources

Anderson, F., *Illustrated History of the Herbals* (New York: Columbia University, 1977).

Ankarloo, B., and S. Clark (eds), *Witchcraft and Magic in Europe: The Period of the Witch Trials* (Philadelphia, PA: University of Pennsylvania Press, 2002).

Arrizabalaga, J., and K. Siena (eds), 'Medical Responses to the "French Diesease" in Europe at the turn of the Sixteenth Century', *Sins of the Flesh: Responding to Sexual Disease in Early Modern Europe.* (Toronto: Center for Reformation and Renaissance Studies, 2005).

Bates, A., *Emblematic Monsters* (Amsterdam: Rodopi, 2005).

Behringer, W. (ed.), *Hexen und Hexenprozesse* (München: Deutscher Taschenbuch Verlag, 1988).

Benzenhöfer, U., and K. Finsterbusch examine anti-Semitism in the works of Paracelsus in their article, 'AntiJudaismus in den Medizinish-Naturwissenschaftlichen und Philosophischen Schriften des Paracelsus', in H. Schott and I. Ziguer (eds), *Paracelsus und Seine Internationale Rezeption in der Frühen Neuzeit* (London: Brill, 1998), pp. 96–109.

Bergengruen, M., *Nachfolge Christi-Nachahmung der Natur* (Hamburg: Meiner, 2007).

Biegger, K., *'De Invocatione Beatae Mariae Virginis': Paracelsus und die Marienverehrung.* (Stuttgart: Steiner, 1990).

—, 'Paracelsus über Maria-Weiblichkeit und Heiligkeit' *Nova Acta Paracelsica*, 14 (2000), pp. 3–18.

Boerlin, P. H., *Leonhard Thurneysser als Auftraggeber: Kunst im Dienste der Selbstdarstellung zwischen Humanismus und Barock.* (Basel and Stuttgart: Birkhäuser, 1976).

Bonjour, E. *Die Universität Basel von den Anfängen bis zur Gegenwart* 1460–1960. (Basel: Helbing und Lichtenhahn, 1971).

Braun, L., 'L'Idée de ‚Matrix' chez Paracelse', in H. Schott and I. Ziguer (eds), *Paracelsus und Seine Internationale Rezeption in der frühen Neuzeit* (Leiden: Brill, 1998).

Brown, P. *The Body and Society: Men, Women, and Sexual Renunciation in Early Christianity.* (New York: Columbia University Press, 1988).

Bullough, V. *The Development of Medicine as a Profession* (New York: Hafner, 1966).

Bynum, C. W. *Wonderful Blood: Theology and Practice in Late Medieval Northern Germany and Beyond* (Philadelphia, PA: University of Pennsylvania Press, 2007).

—, *Fragmentation and Redemption: Essays on Gender and the Human Body in Medieval Religion* (New York: Zone Books, 1991).

—, *Jesus as Mother: Studies in the Spirituality of the High Middle Ages.* (Berkeley, CA: University of California Press, 1982).

Cadden, J. *Meaning of Sex Difference in the Middle Ages: Medicine, Science and Culture* (Cambridge: Cambridge University Press, 1993).

Canning, K., 'Body as Method? Reflections on the Place of the Body in Gender History' *Gender and History* 11;3 (1999), pp. 499–513.

Carlino, A., *Paper Bodies: A Catalogue of Anatomical Fugitive Sheets 1538–1687* trans. N. Arikha (London: Wellcome Institute for the History of Medicine, 1999).

—, *Books of the Body: Anatomical Ritual and Renaissance Learning.* Trans. John Tedeschi and Anne C. Tedeschi (Chicago, IL: University of Chicago Press, 1999).

Cislo, A. E., 'Paracelsus's Conception of Seeds: Rethinking Paracelsus's Ideas of Body and Matter', *Ambix*, 55:3 (November 2008), pp. 274–82.

Cooper, A., *Inventing the Indigenous: Local Knowledge and Natural History in Early Modern Europe* (Cambridge: Cambridge University Press, 2007).

Cramer, T., 'Der Umgang mit dem Wunderbaren in der Natur' *Knowledge, Science and Literature in Early Modern Germany* Gerhild Scholz Williams and Stephan K. Schindler (eds) (Chapel Hill, NC: University of North Carolina Press, 1996).

Crowther, K., 'Creating Adam and Eve: Body, Soul and Gender in Sixteenth Century Germany' (PhD dissertation, Johns Hopkins University, Baltimore, MD 2001)

—, '"Be Fruitful and Multiply": Genesis and Generation in Reformation Germany', *Renaissance Quarterly*, 55 (2002), pp. 905–34.

Daniel, D. T., 'Paracelsus on Baptism and The Acquiring of the Eternal Body', in *Paracelsian Moments*, ed. G. S. Williams and C. D. Gunnoe, Jr (Kirksville: Truman State University, 2002), pp. 117–34.

—, 'Invisible Wombs: Rethinking Paracelsus's Concept of Body and Matter', *Ambix*, 53 (2006), pp. 129–42.

Daston, L. and F. Vidal (eds) *The Moral Authority of Nature.* (Chicago, IL: University of Chicago Press, 2004).

—, Daston, L. and K. Park *Wonders and the Order of Nature 1150–1750* (New York: Zone, 1998)

Debus, A. G., *The Chemical Philosophy: Paracelsian Science and Medicine in the Sixteenth and Seventeenth Centuries* (New York: Science History Publications, 1977).

—, 'Paracelsianism and Chemical Philosophy' in O. P. Grell (ed.) *Paracelsus: The Man and His Reputation His Ideas and Their Transformation* (Leiden, Boston, MA and Köln: Brill, 1998), pp. 225–44

Deichmann, W.B,, D. Henschler, B. Holmstedt and G. Keil, 'What is there that is not Poison?: A Study of Third Defense by Paracelsus'. *Archives of Toxicology*, 58 (1986), pp. 207–13.

DeVun, L. 'The Jesus Hermaphrodite: Science and Sex Difference in Premodern Europe' *Journal of the History of Ideas*, 69:2 (2008), pp. 193–218.

Duden, B., 'The Fetus on the "Farther Shore": Toward a History of the Unborn', in L. M. Morgan and M. W. Michaels, *Fetal Subjects, Feminist Positions* (Philadelphia, PA: University of Pennsylvania Press, 1999), pp. 13–25.

Eamon, W., *Science and the Secrets of Nature: Books of Secrets in Medieval and Early Modern Culture* (Princeton, NJ: Princeton University Press, 1994).

Encarnación, K. R., 'The Proper Uses of Desire: Sex and Procreation in Reformation Anatomical Sheets', in A. L. McClanan and K. R. Encarnación (eds), *The Material Culture of Sex, Procreation, and Marriage in Premodern Europe* (New York: Palgrave, 2002), pp. 221–47.

Ellington, D. S., *From Sacred Body to Angelic Soul: Understanding Mary in Late Medieval and Early Modern Europe* (Washington, DC: Catholic University of America Press, 2001).

Elwood, C., *The Body Broken* (New York: Oxford University Press, 1999).

Engel, G., F. Hassauer, B. Rand and H. (eds) Wunder.*Geschlechterstreit am Beginn der europäischen Moderne.* (Königstein: Ulrike Helmer, 2004).

Fausto-Sterling, A., *Sexing the Body: Gender Politics and the Construction of Sexuality.* (New York: Basic Books, 2000).

Fissell, M., *Vernacular Bodies: The Politics of Reproduction in Early Modern England* (Oxford: Oxford University Press, 2004). Foucault, *The History of Sexuality* trans. Robert Hurley (New York: Vintage Books, 1990), 3 vols.

—, *Discipline and Punish: the Birth of the Prison*, trans. A. Sheridan (New York: Vintage Books, 1995).

French, R. and J. Arrizabalaga, 'Coping with the French Disease: University Practitioner's Strategies and Tactics in the Transition from the Fifteenth to the Sixteenth Century', in R. French, J. Arrizabalaga, A. Cunningham and L. Garcia-Ballester (eds), *Medicine from the Black Death to the French Disease* (Brookfield: Ashgate, 1998), pp. 248–87.

Der Frühparacelsismus, Part 1, ed. Wilhelm Kühlmann and Joachim Telle (Tübingen: Max Niemeyer, 2001).

Der Frühparacelsismus, Part 2, ed. Wilhelm Kühlmann and Joachim Telle (Tübingen: Max Niemeyer, 2004).

Gause, U., *Paracelsus (1493–1541): Genese und Entfaltung seiner frühen Theologie.* (Tübingen: J.C.B. Mohr, 1993).

—, 'On Paracelsus's Epistemology', in O. P. Grell, *Paracelsus: The Man and His Reputation His Ideas and their Transformation* (Leiden: Brill, 1998).

Goldammer, K., *Der Göttliche Magier und die Magierin Natur* (Stuttgart: Franz Steiner, 1991).

Grendler, P. F. 'The Universities of the Renaissance and the Reformation', *Renaissance Quarterly*, 57:1 (Spring 2004), pp. 1–42

Grell, O. P. 'The Acceptable Face of Paracelsianism: The Legacy of Idea Medicinae and the Introduction of Paracelsianism into Early Modern Denmark' O. P. Grell (ed) *Paracelsus: The Man and His Reputation His Ideas and Their Transformation* (Leiden, Boston, Köln: Brill, 1998), pp. 245–67.

Green, M. H., *Making Women's Medicine Masculine: The Rise of Male Authority in Pre-Modern Gynaecology.* (Oxford: Oxford University Press, 2008).

Gunnoe, C. D. 'Paracelsus's Biography Among his Detractors', G. S. Williams and C. D. Gunnoe (eds) *Paracelsian Moments* (Kirksville, MO: Truman State University Press, 2002), pp. 3–17.

Harrison, P., *The Bible, Protestantism, and the Rise of Natural Sci*ence (Cambridge: Cambridge University Press, 1998).

Hirsch, A., *Biographische Lexikon der hervorragenden Ärzte aller Zeiten und Völker*_Vol. 5 (Berlin-Wien: Urban and Schwarzenberg, 1934), pp. 582–3.

Huet, M.-H., *Monstrous Imagination.* (Cambridge, MA: Harvard University Press, 1993).

Jung, C. G., *Psychology and Alchemy*. (New York: Bollingen Series XX, 1953).

Karant-Nunn, S. C., and M. Wiesner-Hanks. *Luther on Women: A Sourcebook.* (Cambridge: Cambridge University Press, 2003).

Keller, H. E. 'Zwo welt in einer haut: Paracelsische "Augenlehr" am Beispiel der Frau'. *Nova Acta Paracelsica,* 14 (2000), pp. 41–78.

—, 'Seeing Macrocosma: Paracelsus's Gendered Epistemology', in G. S. Williams and C. D. Gunnoe, *Paracelsain Moments* (Kirksville, MO: Truman State University Press: 2002), pp. 93–116.

Kristeller, P. O., *Renaissance Thought and It's Sources* (New York: Columbia University Press, 1979).

Laqueur, T. *Making Sex: Body and Gender from the Greeks to Freud* (Cambridge, MA: Harvard University Press, 1990).

Levi, J., 'Paracelsians versus Antiparacelsians: German Doctors against Jewish Physicians in Frankfurt am Main 1618–1645', in H. Schott and I. Ziguer (eds), *Paracelsus und Seine Internationale Rezeption in der Frühen Neuzeit* (Leiden: Brill, 1998), pp. 110–30.

Lundt, B., 'Von Selbstgeburt und Spinnentod. Kontinuität und Kontrast der Genderbilder in erzählenden Texten über Paracelsus'. *Nova Acta Paracelsica,* 14 (2000), pp. 97–132.

Martin, Emily. 'The Egg and the Sperm: How Science has Constructed a Romance Based on Stereotypical Male-Female Roles' *Signs*, 16:3 (1991), pp. 485–501.

McGrath, A., *Reformation Thought* (Oxford: B. Blackwell, 1988).

Merchant, C., *The Death of Nature* (New York: Harper Collins, 1980).

McTavish, L., *Childbrith and the Display of Authority in Early Modern France* (Burlington, VA: Ashgate, 2005).

Midelfort, H.C. E., *A History of Madness in Sixteenth-Century Germany* (Stanford, CA: Stanford University Press, 1999).

—, 'The Anthropological Roots of Paracelsus's Psychiatry', *Medizinhistorisches Journal*, 16 (1981), pp. 67–77.

Millar, I., J Millar and M. Millar (eds), 'Paracelsus', *Cambridge Dictionary of Scientists David* (Cambridge: Cambridge University Press, 2002), p. 281.

Moran, B. T., *Distilling Knowledge: Alchemy, Chemistry, and The Scientific Revolution* (Cambridge, MA: Harvard University Press, 2005).

—, 'Court Authority and Chemical Medicine: Mortiz of Hessen, Johannes Hartmann, and the Origin of Academic Chemiatria', *Bulletin of the History of Medicine*, 63 (1989), pp. 225–46.

—, Der alchemistisch-paracelsische Kreis um den Landgrafen Moritz von Hessen-Kassel (1572–1632)'. *Salzburger Beiträge zur Paracelsusforschung* 25 (1987), pp. 119–45

Multhauf, R., 'Medical Chemistry and "the Paracelsians"' *Bulletin of the History of Medicine,* 28:2 (1954), pp. 101–26.

Newman, W. R. 'Alchemy, Domination and Gender', in N. Koertge (ed), *A House Built on Sand: Exposing Postmodernist Myths about Science* (New York: Oxford University Press, 1998), pp. 216–26.

—, *Atoms and Alchemy: Chymistry & the Experimental Origins of the Scientific Revolution* (Chicago, IL: University of Chicago Press, 2006).

Nummendal, T. *Alchemy and Authority in the Holy Roman Empire.* (Chicago, IL: University of Chicago Press, 2007).

Nutton ,V. 'Medicine at the German Universities, 1348–1500' in A. Cunningham and R. French (eds) *Medicine after the Black Death* (Cambridge: Cambridge University Press,1995), p. 173.

—, Medicine at the German Universities, 1348–1500', *Würzburger medizinhistorische Mitteilungen*, 16 (1997), pp. 173–90.

O'Dowd, M. J. *The History of Medications for Women* (New York: Parthenon, 2001).

Pagel, J. L., *Entwicklung der Medicine in Berlin* (Wiesbaden: Bergman, 1897).

Pagel, W., *Paracelsus; an Introduction to Philosophical Medicine in the Era of the Renaissance* (1958; Basel, New York: S. Karger, 1982).

—, 'Paracelsus and the Neoplatonic and Gnostic Tradition', *Ambix*, 8 (1960), pp. 125–66.

Park, K. *Secrets of Women: Gender, Generation, and the Origins of Human Dissection* (New York: Zone, 2006).

—, 'Nature in Person: Medieval and Renaissance Allegories and Emblems', in L. Daston and F. Vidal (eds), *The Moral Authority of Nature* (Chicago, IL: The University of Chicago Press, 2004), pp. 50–73

—, 'Relics of a Fertile Heart: The "Autopsy" of Clare of Montefalco', in A. L. McClanan and K. Rosoff Encarnación (eds), *The Material Culture of Sex, Procreation, and Marriage in Premodern Europe* (New York: Palgrave, 2002), pp. 115–33.

—, 'Dissecting the Female Body: From Women's Secrets to the Secrets of Nature', in J.Donawerth and A. Seeff (eds) *Crossing Boundaries: Attending to Early Modern Women*. (Newark, DE: University of Delaware Press, 2000), pp. 29–47.

—, 'The Clitoris Rediscovered', in D. Hillman and C. Mazzio (eds), *The Body in Parts*. (New York: Routledge, 1997). pp. 171–94.

—, 'The Life of the Corpse: Division and Dissection in Late Medieval Europe', *The Journal of the History of Medicine and Allied Sciences*, 50 (1995), pp. 111–32.

Partington, J. R. *A History of Chemistry* vol. 2 (London: Macmillan, 1961).

Pörksen, G. 'Microcosma. Der weiblich Leib. Zu einer Theorie der ‚Frauenheilkunde' bei Paracelsus'. *Nova Acta Paracelsica*, 14 (2000), pp. 79–96

Principe, L., (ed) *Chymists and Chymistry: Studies in the History of Alchemy and Early Modern Medicine*. (Sagmore Beach: Chemical Heritage Foundation, 2007).

Rankin, A. 'Duchess, Heal Thyself: Elisabeth of Rochlitz and the Patient's Perspective in Early Modern Germany'. *Bulletin of the History of Medicine*, 82:1 (2008), pp. 109–44.

—, 'Becoming an Expert Practitioner', *ISIS*, 98:1 (2007), pp. 23–53.

Roper, L., *Witch Craze: Terror and Fantasy in Baroque Germany* (New Haven, CT: Yale University Press, 2004).

Rublack, U., 'Pregnancy, Childbirth and the Female Body in Early Modern Germany', *Past and Present*, 150 (February 1996), pp. 84–110.

Rudolph, H. 'Paracelsus' Laien Theologie in traditionsgeschichtlicher Sicht und in ihrer Zuordnung zu Reformation und Katholisher Reform'. in P. Dilg and H. Rudolph (eds.) *Resultate und Disiderate für Paracelsus-Forschung* (Stuttgart: F.Steiner, 1993), pp. 79–97.

Sander, S. '"Aufklärung vor der Aufklärung?": Zum poplulärmedizinischen Werk des Arztes und Bestsellautors Christoph von Hellwig (1663–1721)' *Medizinhistorisches Journal* 34 (1999), pp. 245–308.

Sawday, J., *The Body Emblazoned: Dissection and the Human Body in Renaissance Culture* (London: Routledge, 1995).

Schiebinger, L., 'Skeletons in the Closet: The First Illustrations of the Female Skeleton in Eighteenth-Century Anatomy'. in L. Shiebinger (ed) *Feminism and the Body* (Oxford: Oxford University Press, 2000), pp. 25–52.

Schott, H. and I. Ziguer eds. *Paracelsus und Seine Internationale Rezeption in der Frühen Neuzeit* (Leiden: Brill, 1998).

Shackelford, J. 'Paracelsianism and Patronage in Early Modern Denmark'. in B. T. Moran (ed.), *Patronage and Institutions: Science, Technology, and Medicine at the European Court 1500–1750*. (Rochester, NY: Boydell, 1991), pp. 85–109.

—, *A Philosophical Path for Paracelsian Medicine: The Ideas, Intellectual Context, and Influence of Petrus Severinus, 1540–1602 (Acta historica scientiarum naturalum et medicinalium)* (Copenhagen: Museum Tusculanum Press, 2004).

Shatzmiller, J. *Jews, Medicine and Medieval Society* (Berkeley, CA: University of California Press, 1994).

Schmidt, S.K. 'Art – A User's Guide: Interactive and Sculptural Printmaking in the Renaissance' (PhD dissertation Yale University May 2006).

Sirasi, N. *The Clock and the Mirror* (Princeton, NJ: Princeton University Press, 1997).

—, *Medieval and Early Renaissance Medicine: An Introduction to Knowledge and Practice*. (Chicago, IL: University of Chicago Press, 1990).

Spinks, J., 'Wondrous Monsters: Representing Conjoined Twins in Early Sixteenth-Century German Broadsides', *Parergon*, 22:2 (2005), pp. 78–112.

—, *Monstrous Births and Visual Culture in Sixteenth-Century Germany* (London: Pickering and Chatto, 2009).

Spitzer, G. *...und die Spree führt Gold. Leonhard Thurnhesser zum Thurn Astrologo-Alchimist-Arzt und Drucker im Berlin des 16. Jahrhunderts* (Berlin: Staatsbibliothek zu Berlin Preussischer Kulturbesitz, 1996).

Steiger, J. A., *Medizinische Theologie: Christus medicus und theologia medicinalis bei Martin Luther und im Luthertum der Barockzeit.* (Leiden: Brill, 2005).

Stolberg, M. 'Critiques and Conventions: A Woman Down to Her Bones', *ISIS*, 94:2 (2003), pp. 274–99.

Walton, M., 'Anthonius Margaritha – Honest Reporter?', *Sixteenth Century Journal*, 36:1 (2005), pp. 129–41.

Wear, A., 'Medicine in Early Modern Europe 1500–1700', in L. Conrad (ed.) *The Western Medical Tradition 800 B.C. to 1800* (Cambridge: Cambridge University Press, 1995), pp. 215–370.

Webster, C., *Paracelsus: Medicine, Magic and Mission at the End of Time* (New Haven, CT: Yale University Press, 2008)

—, 'Medicine as Popular Protest in Medicine and the Reformation'. in O.P. Grell and A. Cunningham (eds) *Medicine and the Reformation* (London: Routledge, 1993), pp. 57–77

—,*From Paracelsus to Newton: Magic and the Making of Modern Science*. Mineola: Dover, 1982

Weeks, A., *Paracelsus: Theophrastus Bombastus von Hohenheim 1493–1541: Essential Theoretical Writings* (Leiden: Brill, 2008).

—, *Paracelsus: Speculative Theory and the Crisis of the Early Reformation* (Albany, NY: State University of New York Press, 1997).

Wiesner, M., *The Art of Midwifery: Early Modern Europe* (London: Routledge, 1993).

—, *Women and Gender in Early Modern Europe* (Cambridge: Cambridge University Press, 2000).

Williams, G. H., *Radical Reformation* (Kirksville, MO: Sixteenth Century Journal Publshers, 1992).

Williams, G. S.. *Defining Dominion* (Ann Arbor, MI: University of Michigan Press, 1995).

—, 'Die dritte Kreatur. Das Frauenbild in den Schriften von Paracelsus (1491–1543)', in I. Bennewitz (ed.), *Der frauwen bouch: Versuche zu einer feministischen Germanistik* (Göppingen: Kümmerle, 1989), pp. 351–71.

INDEX

For Product Safety Concerns and Information please contact our EU
representative GPSR@taylorandfrancis.com
Taylor & Francis Verlag GmbH, Kaufingerstraße 24, 80331 München, Germany

www.ingramcontent.com/pod-product-compliance
Ingram Content Group UK Ltd.
Pitfield, Milton Keynes, MK11 3LW, UK
UKHW011017180926
475701UK00004B/788
9781138663824